Topology Optimization of Compliant Mechanisms

Xianmin Zhang · Benliang Zhu

Topology Optimization of Compliant Mechanisms

Springer

Xianmin Zhang
Guangdong Key Laboratory of Precision Equipment and Manufacturing Technology
School of Mechanical and Automotive Engineering, South China University of Technology
Guangzhou, Guangdong
China

Benliang Zhu
Guangdong Key Laboratory of Precision Equipment and Manufacturing Technology
School of Mechanical and Automotive Engineering, South China University of Technology
Guangzhou, Guangdong
China

ISBN 978-981-13-4415-2 ISBN 978-981-13-0432-3 (eBook)
https://doi.org/10.1007/978-981-13-0432-3

Printed on acid-free paper

This Springer imprint is published by the registered company Springer Nature Singapore Pte Ltd.
part of Springer Nature
The registered company address is: 152 Beach Road, #21-01/04 Gateway East, Singapore 189721, Singapore

Preface

Compliant mechanisms are flexible structures which can generate the desired motion by undergoing elastic deformation instead of through rigid linkage/joints as in rigid body mechanisms. It is suitable to be used in the fields of precision engineering especially in the fields of micro-manipulation and micro-positioning. The suitability is mainly due to the monolithic nature of compliant mechanisms which in turn gives rise to advantage such as no need for assemble, no friction/wear (and hence no need for lubrication), and no backlash.

Over the past decades, methods for the design of compliant mechanisms have been developed and can be categorized into two general areas, pseudo-rigid body methods and topology design methods. Compared with the pseudo-rigid body methods, topology design methods have the advantages that the designer need not begin with a known rigid-link mechanism, and the resulting mechanism configurations are optimal in terms of the input force–output deflection behavior.

This work attempts to provide an unified presentation of designing compliant mechanisms by using the topology optimization methods. This book is written for academic researchers with an interest in topology optimization and compliant mechanisms. It contains five chapters. The basic theory of topology optimization of compliant mechanisms is demonstrated in Chap. 1. Flexure hinges are the most important components of compliant mechanisms, various cross-sectional profiles of flexure hinges, such as the right circular, elliptical, corner-filleted, hyperbolic, parabolic, v-shape, etc., have been widely used in various micro-manipulation and micro-positioning systems. The merits and faults of these known shapes shown that these shapes may not be the best topologies of flexure hinges. Chapter 2 provides a new topology optimization method to find some new flexure hinges. Compared with the lumped compliant mechanisms, the main advantage of distributed compliant mechanisms is that there are no flexible hinges on designed mechanisms, which can greatly reduce stress concentration and raise fatigue life. Therefore, distributed compliant mechanisms are preferable in real-world applications especially suitable for micro-electromechanical systems. How to obtain distributed compliant mechanisms by topology optimization method is presented in Chap. 3. Compliant mechanisms with multiple inputs and multiple outputs are widely used in the fields of micro-positioning and micro-manipulation. In Chap. 4, we developed a multiple objective function for topology design of multiple inputs and outputs compliant mechanisms based on one design domain. Regarding the compliant limbs of compliant parallel mechanisms as separate design domains, a new Jacobian-based topology optimization method for compliant parallel mechanisms is also presented in this chapter. Some extensions are presented in the last chapter, which include the strategy to avoid re-initialization in level set method and to improve the computational efficiency, design of compliant mechanisms considering geometrical nonlinearity, and the reliability-based design of compliant mechanisms. The authors hope that the book is helpful to researches in the field of analysis and design of compliant mechanisms and the related areas.

We wish to thank the support from the National Natural Science Foundation of China (Grant Nos. 50825504, 91223201, 50775073, 50375051, 51605166, U1609206) and the Natural Science Foundation of Guangdong Province (Grant Nos. S2013030013355, 050064, 036548). The first author would like to express his sincere thanks to his students, Gaofei Ouyang, Zhaokun Li, Yongqing Fu, Jinqing Zhan, Dongmei Li, Qiliang Wang, Yongjian Chen, Shaofang Liu, and Kai Hu. Their efforts contribute a lot to this book. Also, we would like to acknowledge Nianfeng Wang, Zhicheng Qiu, Zhong Chen, Yanjiang Huang, and Lixin Yang for their valuable suggestions in the field of topology optimization in the weekly seminars.

Special thanks go to Min Liu and Mohui Jin for providing the drafts of Chaps. 2 and 4, respectively. We surely are also indebted to our colleagues at South China University of Technology for their support, help, and care.

As aforementioned, there are tremendous high-quality papers in the research field of topology optimization of compliant mechanisms. Part of them has been cited at the end of each appropriate chapter. The reader can benefit from referring to these articles and the references therein. The authors of these articles are also greatly appreciated.

Guangzhou, China
December 2017

Xianmin Zhang
Benliang Zhu

The original version of the book was revised: Incorrect first author name has been corrected. The erratum to the book is available at https://doi.org/10.1007/978-981-13-0432-3_6

Contents

Chapter 1
Introduction to Compliant Mechanisms and Design Methods

Abstract "Compliant mechanisms" often refers to monolithic or jointless structures that transfer an input force or displacement to another point through elastic deformation. The intrinsic advantages associated with compliant mechanisms include their lack of a need for lubrication and assembly and their high accuracy. These advantages make compliant mechanisms extremely suitable for precision engineering applications that require nanometer or even subnanometer positioning accuracy. This introductory chapter briefly presents compliant mechanisms, their applications and the associated design approaches.

1.1 Concept of a Compliant Mechanism

Mechanisms that are capable of transmitting motion, force and energy play a key role in machinery. Based on their intrinsic ways of transferring motion and force, mechanisms can be divided into two categories: rigid-body mechanisms and compliant mechanisms.

Rigid-body mechanisms generally consist of moving components, linkages, friction devices, structural components and a variety of specialized machine elements. These components are collected to achieve a specific function. When the movement of a rigid-body mechanism is analyzed, its links are modeled as rigid bodies. Therefore, the relative motion between points in two connected links is considered a result of the kinematic pair that joins them [48, 93]. Examples of the rigid-body mechanisms can be easily found. One example is the Baxter robot [120] shown in Fig. 1.1. It is designed with seven degrees of freedom (7 DOFs) on each arm to provide kinematic redundancy.

Unlike rigid-body mechanisms, compliant mechanisms transmit motion, force or energy by elastic deformation of the materials that comprise them [35]. A four-bar rigid-body linkage and a four-bar compliant mechanism are shown in Fig. 1.2a and b. Their working states, shown in Fig. 1.2 with dashed lines, clearly demonstrate the differences in their work principles.

Compliant mechanisms offer several advantages over with rigid body mechanisms, such as fewer assembly process requirements, backlash elimination, and no

X. Zhang and B. Zhu, *Topology Optimization of Compliant Mechanisms*,
https://doi.org/10.1007/978-981-13-0432-3_1

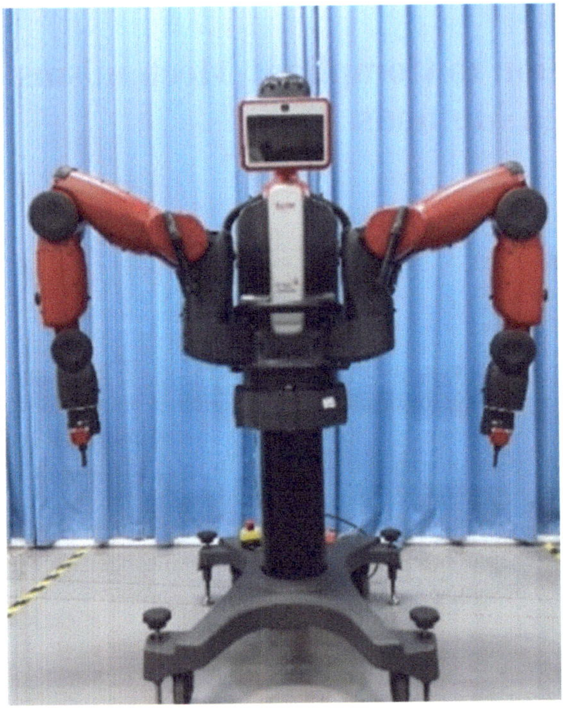

Fig. 1.1 A Baxter robot

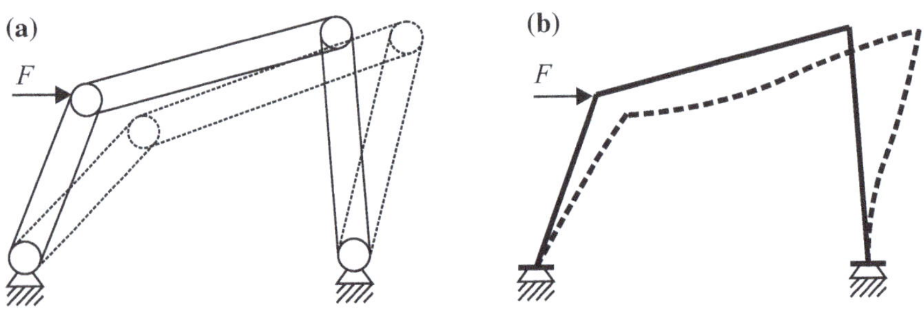

Fig. 1.2 Schematic of a four-bar planar rigid mechanism (**a**) and a four-bar planar compliant mechanism (**b**); the dashed lines represent deformed configurations

need for lubrication. As a consequence, compliant mechanisms have largely been used in a variety of high precision applications, such as precision positioning stages and nano-handling devices in MEMS (micro-electro-mechanical systems).

1.2 Applications of Compliant Mechanisms

The idea of creating motion using flexible structures can be traced back to the very early stage of human history in which the bow and arrow were used for hunting and war. Although the history of using compliant mechanisms or structures is relatively long, records of ancient compliant mechanisms are very rare. The design procedure was based on handicraft rather than technology, which to a large extent depends on the designer's cleverness and talent. For example, a compliant catapult was developed centuries ago by the great artist Leonardo da Vinci [35].

The situation has changed substantially since the 19th century, especially during the last decades. With the rapid development of technology and science, compliant mechanisms or products that rely on flexible members have become involved in increasing numbers of application areas. Among these areas, the two most promising applications are precision positioning stages and MEMS.

A precision positioning stage is a device that can perform very small motions with very high positioning accuracy [47, 48, 102]. The rapid development of high-precision positioning stages is mainly because of a steady need for increased accuracy and precision in production machinery. Typical applications are chip assembly in the semiconductor industry and handling and manipulation at the micro- and nanoscales.

The piezoelectric (PZT) actuation integrated with compliant mechanisms is often regarded as the technology with the greatest promise of high accuracy. There is world-wide interest in creating various high-precision positioning stages with between one and six degrees of freedom (DOFs). Figure 1.3 shows some with different numbers of DOFs. There are several papers presenting design methods and applications that incorporate compliant mechanisms into specific designs. A few them are mentioned in what follows.

The earliest reports on compliant mechanism-based positioning stages date to the 1970s. The created stages consist of piezoelectric actuators and flexure hinges which can produce linear motions [30, 77]. For example, Scire and Teague [77] developed a single-DOF compliant stage by combining a PZT with a flexure-pivoted lever arm operating as a displacement amplifier. However, it should be noted that such compliant mechanisms were only able to generate linear motion, which limited their applications. The situation has changed since the 1980s, when multi-DOF micro-positioning stages that can produce more complex motions were developed [46, 95, 98, 100, 101, 115]. Examples can be found in, e.g., [46], where a three-DOF parallel actuated manipulator was developed, and [115], where an XY positioning stage was developed for end-point sensing. Since the 1990s, tremendous compliant mechanism-based positioning stages have been developed alongside actuation strategies, control methods and methods for achieving large strokes. Please refer to a previous review article [66] and the references therein for more information about this area.

The other application of compliant mechanisms is in the area of the so-called MEMS. MEMS technology has been regarded as one of the greatest technologies in 20th century. It all began with a paper published by Charles Smith of Bell Labs in 1954 that described stress-sensitive effects in silicon and germanium. Over the

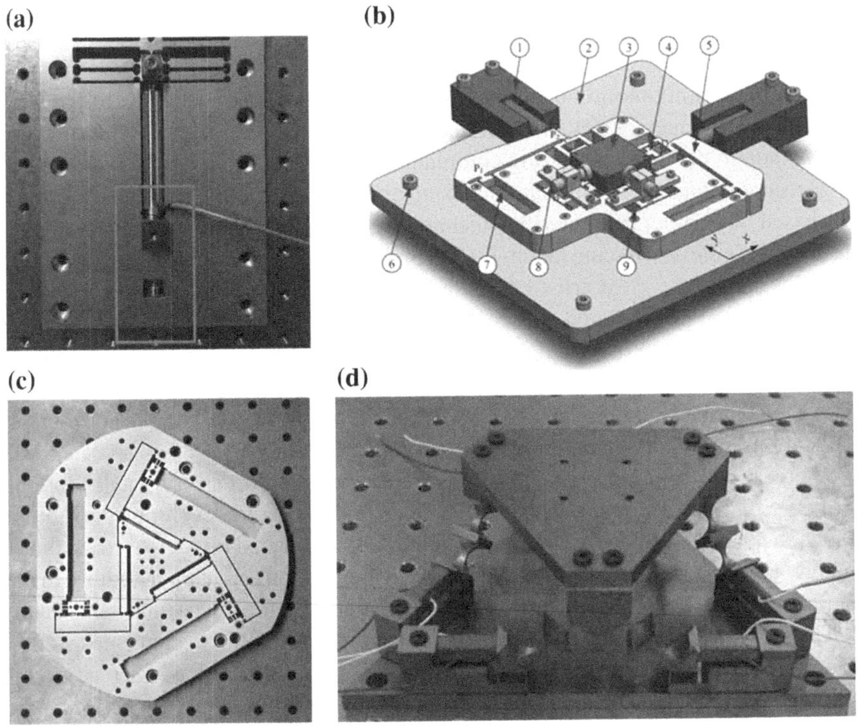

Fig. 1.3 Precision positioning stages that are designed using compliant mechanisms with different numbers of DOFs: **a** 1 DOF [99], **b** 2 DOFs [90], **c** 3 DOFs [98], and **d** 6 DOFs [107]

following decades, MEMS products, such as automobile airbag systems, display systems and sensors, have given rise to a multi-billion market [14].

MEMS, as their name implies, are very small. They are often made up of components with sizes from 1 to 100 μm, and MEMS devices generally range in size from 20 μm to 1 mm [92]. For such small scales, mechanisms connected by rigid movement pairs are very difficult or even impossible to fabricate using current micromachining technologies, which are primarily based on deposition and etching. This inherent property inevitably leads to significant challenges in designing MEMS structures.

Being monolithic, compliant mechanisms therefore become a possible or, sometimes, the only choice for MEMS structures. The micro-gripper provides an example. The micro-gripper has been widely used in MEMS for manipulating small objects, such as CNTs (carbon nanotubes) [5, 19, 26, 73]. Regardless of the actuator used, the structure of the gripper itself, e.g., the two beams/fingers [80], is often designed based on the designers' intuition or inspiration. This means that the design of a compliant mechanism-based MEMES structure is a handcraft rather than a technology. However, the situation has changed significantly over the past several decades because some design optimization algorithms have been employed to create conceptual designs of MEMS devices (Fig. 1.4). This also paves the way for rapid increases in the use of compliant mechanisms in MEMS.

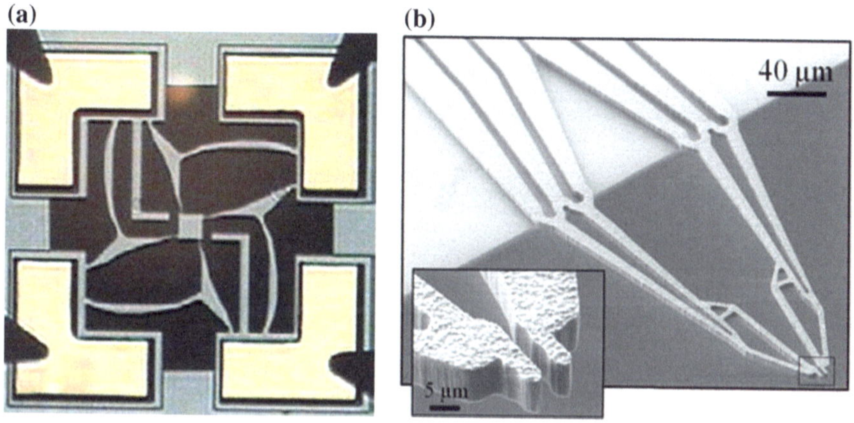

Fig. 1.4 MEMS devices designed using compliant mechanisms: **a** a rotator [34] and **b** a compliant micro gripper [73]

1.3 Approaches to Designing Compliant Mechanisms

Over the past several decades, compliant mechanisms have been extensively studied. Although the methodology for designing compliant mechanisms remains an open area, several approaches to the systemic design of compliant mechanisms have been developed. One of the first approaches used for designing compliant mechanisms was based on converting analogous rigid-body mechanisms into compliant mechanisms; this is well known as the kinematics-based approach (also known as the pseudo-rigid-body model). A second approach to designing compliant mechanisms involves the use of structural topology optimization. The following sections briefly illustrate the details of these two approaches.

1.3.1 Kinematics-Based Approach

In this approach, the compliant mechanism is derived from a known rigid-body mechanism. The kinematics-based approach is based on traditional rigid-body kinematics wherein a rigid-body mechanism is first synthesized. Then, one can revise the rigid-body mechanism design by introducing flexible members. Therefore, a quick way to reach a viable design is to use of rigid-body replacement synthesis. Figure 1.5 illustrates a typical design process for a compliant 3RRR positioning stage using the kinematics-based approach. The kinematics-based approach is attractive since it actually provides an available way to quickly explore a variety of compliant mechanism configurations.

Compared with rigid-body components that can have the ability to undergo continuous revolution (e.g., a hinge joint), flexible components can only undergo limited

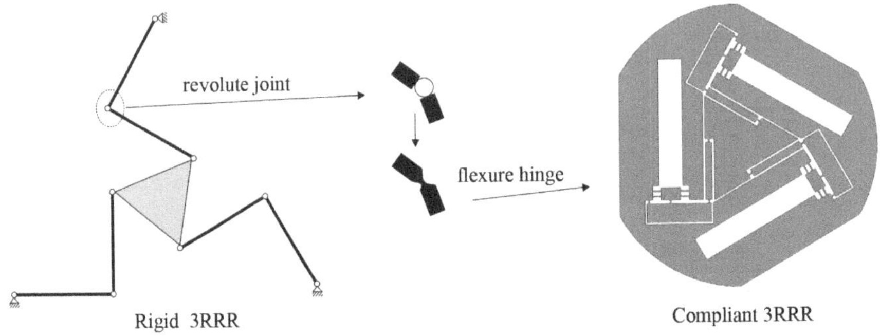

Fig. 1.5 Schematic of the process of designing compliant mechanisms using the kinematics-based approach. A 3-RRR parallel compliant mechanism is considered

motion due to the limit stress of the material. After obtaining the configuration of a compliant mechanism, such as that shown in Fig. 1.5, the pseudo-rigid-body model [35, 36] can be adopted to analyze the motion of the mechanism. The underlying idea of the pseudo-rigid-body model is to model the deflection of a flexible member using a rigid-body component that has equivalent force-deflection characteristics. This basic idea can be illustrated using Fig. 1.6, where a short flexural pivot (cantilever) is considered. The corresponding pseudo-rigid-body model is shown in Fig. 1.6b. The motion of the cantilever can be modeled using two rigid links jointed at a pin joint, which is called the characteristic pivot. The location of the characteristic pivot can be measured from the end of the cantilever as a fraction of the cantilever's length. The distance γl is called the characteristic radius, and γ is called the characteristic radius factor. A torsional spring is added at the pivot to represent the resistance of the cantilever's deflection.

The pseudo-rigid-body model actually establishes a bridge that connects the rigid-body mechanism theory and the compliant mechanism theory; therefore, it is particularly useful in the design of compliant mechanisms. The relationship between a compliant mechanism and its corresponding pseudo-rigid-body model has been extensively studied. Some of the important results can be found in [39, 106, 116, 131]. The validity of the pseudo-rigid-body model for precisely predicting the motion of compliant mechanisms has been verified using both finite element analysis and experiments. Please refer to [35] for more details.

Although the kinematics-based approach has been successfully used in designing compliant mechanisms, it sometimes requires a good deal of intuition and involvement on the part of the designers. Furthermore, compliant mechanisms that are designed using the kinematics-based approach often have lumped compliance which means that, when the mechanisms are loaded, only the material near the hinges contributes to the deflection at the output ports. For this reason, the obtained compliant mechanisms are very difficult to miniaturize so that they can be used on the micro- or nano-scale. To design devices that are suitable for MEMS, topology optimization methods are preferred because they generate distributed compliant mechanisms.

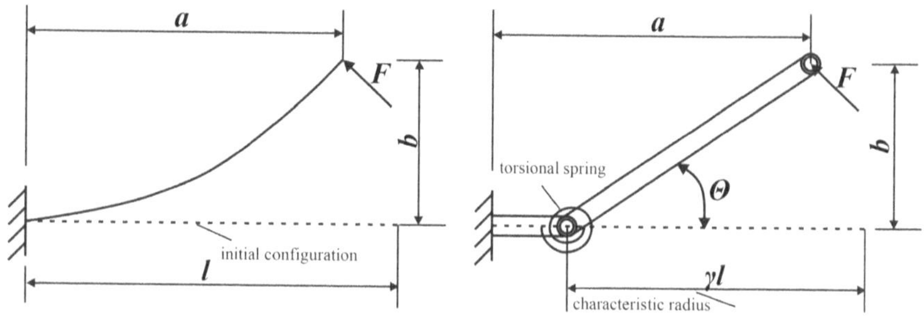

Fig. 1.6 Schematic of the pseudo-rigid-body method [35]

Table 1.1 Compliant mechanisms, associated design methods and typical applications

Compliant mechanism	Design method	Typical application area
Lumped	Pseudo-rigid body method	Precision positioning stages
Distributed	Topology optimization	MEMS devices

A distributed compliant mechanism is capable of producing distributed compliance through the elastic deformation of the entire structure. Therefore, we can use Table 1.1 to summarize the categories of the compliant mechanisms, their corresponding design methods and typical application areas.

1.3.2 Topology Optimization Approach

Methods for optimizing structures can generally be categorized according to three design variables: size, shape and topology optimization. For size optimization, the considered design variables relate to the geometric parameters of the structure. An example is shown in Fig. 1.7a, where a block with a circular slot is considered. The location of the slot is fixed, and the size of the slot is determined by the radius r, which is the size design variable. One can change the radius r of the slot to reduce the weight of the block. However, the geometry of the structure does not change. For shape optimization, the design variable is related to the geometry of the structure. Take the block as an example again. The design variable r in Fig. 1.7a only determines the size of the circle. For some cases, the circular shape may be not a good choice. One may consider other shapes, such as ovals, as shown in Fig. 1.7b. To control the shape of the slot, one can use parametric curves, e.g., a closed Bézier curve, to represent the boundary of the slot. In this case, the locations of the control points can be considered the shape design variables.

Topology optimization is concerned with seeking the optimum distribution of material in a given design domain that minimizes a given cost function while satisfying a series of constraints [11, 13]. Unlike the shape optimization variable that

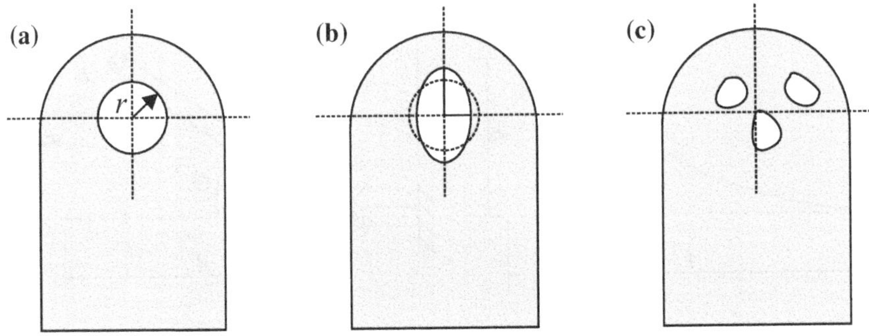

Fig. 1.7 Size (**a**), shape (**b**) and topology (**c**) optimization

represents changes in structural geometry, topology design determines the structure's layout. Shape optimization can change the shape of the slot but cannot introduce a new one. The unique feature of topology optimization is that it can determine whether the slot can be removed or an additional slot is required, as shown in Fig. 1.7c. Early developments in structural topology optimization focused on truss structures [55, 56]. The basic idea of this method is that one uses truss or beam elements to connect a given set of points in a space to identify the layout that supports the largest load. Recent developments in structural topology optimization are strongly related to continuum finite element analysis. Starting with the seminal work of Bendsøe and Kikuchi [11], numerical methods for continuum structural topology optimization have been extensively investigated, and several approaches have been developed.

Topology optimization of compliant mechanisms can be performed based on either continuum or truss discretizations. Each type has its advantages and disadvantages. The following sections briefly introduce the underlying idea of each method and its characteristics.

1.3.2.1 Ground Structure Approach

The ground structure approach uses an exhaustive set of truss or beam/frame elements in the design domain. The individual cross-sectional area is defined as a design variable. When the cross sectional area of an element reaches zero or approaches the lower bound of the design variable in the optimization design process, that element is removed. Thus, when the optimization procedure has converged, some elements have been removed from the original exhaustive set. The remaining elements define the topology of the optimal structure [23, 69, 117].

If beam members are used to represent the design domain, they exhibit bending moments as well as axial loads, and the joints are usually assumed to be rigidly connected at both ends so bending moments can be transferred from one element to another. This enables a better description of the deformation of the compliant mechanism; therefore, the ground structure method is very suitable for the optimal

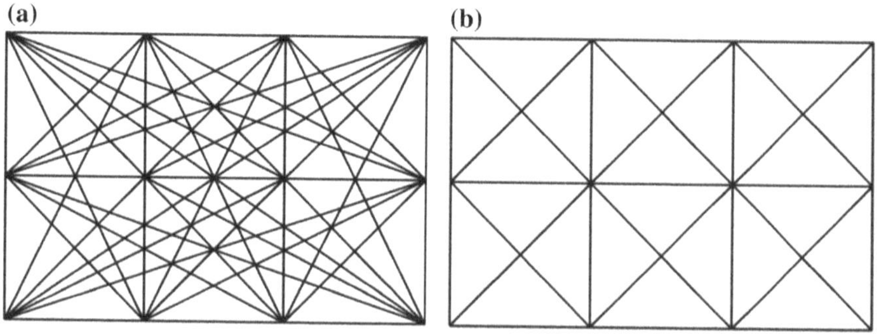

Fig. 1.8 Ground structure approach with **a** a fully connected ground structure and **b** a reduced ground structure

design of a compliant mechanism. Two types of ground structure are typically used with truss-type members: a fully connected ground structure and a reduced ground structure, as shown in Fig. 1.8. A reduced ground structure of frame elements is usually implemented when designing the topology of a compliant mechanism because fully connected ground structures have overlapping elements that are difficult to manufacture.

The primary work on synthesizing the topology of a compliant mechanisms using the ground structure method is presented in [45], where the flexibility and the stiffness were considered simultaneously by using a weighted sum. The output deformation in a given direction was maximized using an objective function, and the strain energy was minimized to prevent the mechanism from succumbing to external loads. However, it was difficult to preset the weighting factor. Therefore, an improvement was proposed by Frecker et al. [28], who used the ratio method, which involves optimizing the ratio of the output displacement and the strain energy.

Plenty of studies have focused on the optimizing the topology of compliant mechanisms using the ground structure approach. Some of them are listed below. Saxena [74] and Joo [43] proposed multi-objective topology optimization in compliant mechanisms with line beam elements using the ground structure approach. Saxena [75] adopted stress constraints in addition to maximizing the output displacement for topological design in compliant mechanisms. Saxena [76] and Joo [42] employed geometric nonlinear finite element analysis to synthesize compliant mechanisms with large deformations. Mankame [57] synthesized the topology of electrothermal compliant mechanisms using line elements. Kinoshita [44] proposed a sequential optimization approach to topology synthesis in bistable compliant mechanisms. Rai [68] presented a new method for the synthesis of planar linkages with rigid and deformable members that may take the form of a rigid body, fully compliant or partially compliant mechanism.

1.3.2.2 Continuum Approaches

The other way is to model the design domain of the topology optimization problem using the continuum approach. In this approach, the candidate design domain is often discretized into quadrilateral finite elements, and then, the material property of each element is controlled. Compared with the ground structure model, the structural model can closely represent a continuum using a fine mesh. During the topology optimization process, when it is necessary to insert a hole (remove a region), one can let the material property values of the elements in this area approach zero. Therefore, the material property design variables can be treated as topology design variables.

For a topology design problem, the optimal solution is a structure that is composed of only one material and voids, which means that a 0–1 decision is used to determine whether the material should be placed at a point within the design domain. In reality, for a 2D problem, the discretization of the design domain typically requires more than 1000 elements. Because of the size of this discrete problem, it is very difficult to use non-gradient-based optimization approaches to solve the design problem [85]. The most common method is to use a continuous variable formulation so it is possible to use gradient-based searches. For example, if we take the Young's modulus E of each element as a design variable, then empty and solid elements have $E = 0$ and $E = 1$, respectively. Relaxation enables the Young's modulus of an element to vary continuously from 0 to 1.

However, relaxation often leads to the appearance of intermediate material states, i.e., Young's moduli such that $0 < E < 1$. Such a result is very far from the 0–1 solution required by topology optimization because regions with intermediate values cause problems in manufacturing. To overcome the intermediate material problem, one may penalize the elements with intermediate Young's moduli [81].

The homogenization method The first continuum topology optimization method that was introduced for designing compliant mechanisms is the so-called homogenization method [11]. This approach converts complex structural topology optimization problems to size optimization problems by introducing a material density function in each element that is composed of an infinite number of periodically distributed holes. Homogenization theory is used to determine the mechanical properties of the materials. The methods of introducing microstructures can be divided into two categories: methods based on rank laminate composites and methods based on microcells with internal voids. The homogenization equation can be analytically solved for the former, whereas numerical methods are often adopted to solve the homogenization equation for the later. The homogenization method has been used to solve various topology optimization problems in many studies. One advantage of the homogenization method is that it can provide mathematical bounds on the theoretical structural performance [32].

The homogenization method was used to design compliant mechanisms in [4]. It was Ananthasuresh who first extended the homogenization approach to the design of compliant mechanisms. The results, however, seem to be a mean compliance design rather than a compliant mechanism design since the obtained mechanisms do not have sufficient flexibility. Therefore, Nishiwaki [61] proposed a homogenization-based topology optimization method that includes flexibility for the design of compliant mechanisms. This method introduced a mutual mean compliance to represent the flexibility and built it into a multi-objective function. Several numerical examples were used to demonstrate that the concept of flexible structure design can be used to design compliant mechanisms. Currently, papers about using the homogenization method to design compliant mechanisms are relatively rare because the homogenization method often requires more complex calculations than other established topology optimization methods.

Starting with the homogenization method, numerical methods for topology optimization have been investigated extensively. Several methods have been developed, including the SIMP (solid isotropic microstructure with penalization) method [12, 70, 81], the ESO (evolutionary structural optimization) method [110–112], the level set method [3, 96, 126], the phase field methods [89], and the topological derivatives method [24, 62]. The following sections briefly introduce the SIMP, ESO, and level set methods and their use in designing compliant mechanisms.

The SIMP method The SIMP method has been generally accepted and has reached the stage of application in industrial software due to its computational efficiency and conceptual simplicity [13, 72]. The SIMP method takes the density ρ of an element as the design variable to control whether it should be made up of void or solid material. The element density ρ_e is calculated using

$$\rho_e = x^e \rho_0, \tag{1.1}$$

where ρ_0 is the density of the reference material and $0 < x^e_{\text{min}} < x^e \leq 1$. If the relative element density x^e is equal to 1, then $\rho_e = \rho_0$, which means that the element consists of solid material. If the relative element density x^e is equal to x^e_{min}, which is a very small number to avoid numerical singularities, then the element is empty. The SIMP method applies a penalty when calculating the element stiffness k^e to avoid intermediate densities;

$$\mathbf{k}^e = (x^e)^p \mathbf{k}^0, \tag{1.2}$$

where $\mathbf{k}^0$ is the stiffness matrix of an element consisting of solid material and $p > 1$ is a free parameter to penalize intermediate densities; the result is approximately a $0 - 1$ design. Usually, $p \geq 3$ is required.

Several mathematical methods can be employed to solve the topology optimization problem, including sequential linear programming [121, 122] and sequential convex programming [10, 27]. Methods involving optimality criteria (OC) are particularly efficient for problems with many variables and few constraints [83]. For optimization problems that have several constraints, sequential linear programming is often employed. One of the most popular optimizers is the MATLAB implementation of the method of moving asymptotes, which is freely available for research purposes from Krister Svanberg [88].

When using the above-mentioned continuum topology optimization methods, special attention must be paid to numerical instabilities, such as mesh dependency and checkerboard patterns. Mesh dependency refers to the fact that different mesh sizes or discretizations may lead to different optimal solutions. Checkerboard patterns occur when the obtained optimal topology contains regions of alternating solid and void elements in a checkerboard-like pattern. Methods for dealing with these numerical instabilities have been extensively studied [21, 41, 71, 86]. Among the established methods, substantial computational experience has demonstrated that filtering is a highly efficient way to overcome mesh dependency and checkerboard patterns [84]. Filtering requires very little CPU time and is simple to use since it does not involve any extra constraints. For more information on methods of handling numerical instabilities, the reader is referred to [84] and the references therein.

The primary study of designing compliant mechanisms using the SIMP method can be found in [82], where the geometric advantage (the ratio of the output and input displacements) of the mechanism is taken as the objective function. To date, several compliant mechanism topology design formulations that use the SIMP method have been developed. The most common one is summarized here. The problem of designing a compliant mechanism with the optimal topology is illustrated in Fig. 1.9. At the input port i, an input force F_{in} with displacement magnitude u_{in} is applied. At the output port o, an output displacement with magnitude u_{out} is desired. Specifically, the input spring k_{in} and the output spring k_{out} represent an actuator and a workpiece, respectively. Therefore, the design problem is to maximize the output displacement u_{out} subject to a material mass constraint $Vol = \sum_{e=1}^{N} \rho_e v_e \leq V^*$, which can be expressed as follows:

$$
\begin{aligned}
\min: \quad & -u_{out} \\
\text{s.t.}: \quad & \sum_{e=1}^{N} \rho_e v_e \leq V^* \\
& 0 < \rho_{\min} \leq \rho \leq 1, \quad e = 1, 2, \ldots, N,
\end{aligned}
\tag{1.3}
$$

where v_e is the volume of element e.

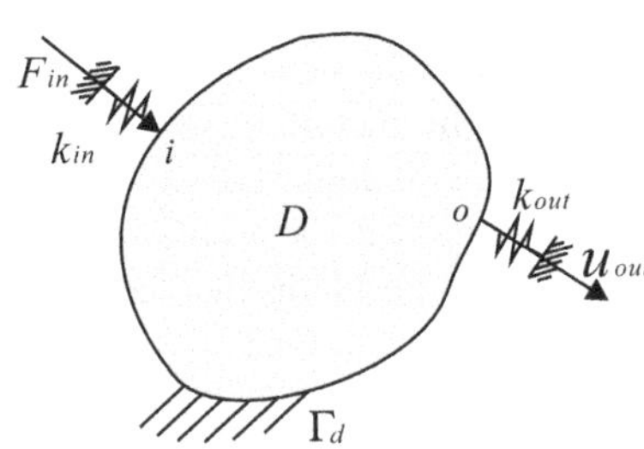

Fig. 1.9 A general design domain for compliant mechanisms when using continuum topology optimization approaches

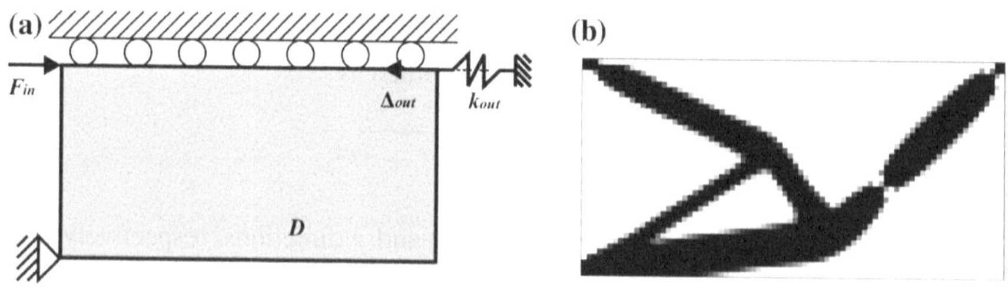

Fig. 1.10 A displacement inverter obtained using the SIMP method to solve Eq. (1.3)

A design example using Eq. (1.3) is shown in Fig. 1.10, where a displacement inverter is considered. The design goal is to obtain a mechanism that can transfer the input force to the opposite direction. This problem was solved using a 104-line MATLAB program for compliant mechanism synthesis obtained from [13]. The parameters used are `topm(80, 40, 0.3, 3.0, 1.5)` and the spring stiffnesses k_{in} and k_{out} are both set to 0.01. One can surely specify different values for those parameters to obtain different designs.

Note that the obtained compliant mechanism shown in Fig. 1.10 contains areas that act like revolute joints in rigid-body linkages. These areas are widely known as de facto hinges [125]. De facto hinges are generally undesirable since large stresses may occur in these areas when the compliant mechanism is deformed. Furthermore, compliant mechanisms with de facto hinges are very difficult to miniaturize, especially on a small scale. De facto hinges have been a point of contention in the literature, and many efforts have been made to avoid them. This topic is also addressed in this book.

Other compliant mechanism design issues that have been addressed include geometric nonlinearity [17, 49], multiple inputs and multiple outputs [50, 119] and incorporation of nonmechanical actuation and contact [20, 25]. In addition, synthesis with multimaterial topology optimization has been addressed [83, 118]. For example, Sigmund [83] presented a method of using topology optimization to design multiphysics actuators and electrothermomechanical systems.

The ESO method The evolutionary structural optimization (ESO) method was developed based on the simple idea of gradually removing inefficient material from a structure to obtain the optimal structure. This method essentially relates to the classical fully stressed design. In an ideal structure, the stress at every point is near the same safe level, which leads to a rejection criterion based on the local stress level. Any material under low stress is assumed to be inefficient and therefore, can be removed [9, 91, 108, 112].

As other topology optimization methods are, the ESO method is closely associated with the finite element method. The design domain is firstly divided into many small elements. Therefore, material can be deleted from the structure by deleting elements from the finite element model. There are several ways to measure the stress in each element, and the von Mises stress is one of the most commonly adopted criteria. For plane stress problems, the von Mises stress is defined as

$$\sigma_v = \sqrt{\sigma_x^2 - \sigma_x\sigma_y + \sigma_y^2 + 3\tau_{xy}^2}, \tag{1.4}$$

where σ_x and σ_y are the normal stresses in the x and y directions, respectively, and τ_{xy} is the shear stress.

To complete the optimization process, one must identify the conditions under which an element needs to be deleted. This is done by comparing the von Mises stress of the element σ_v^e to the maximum von Mises stress of the whole structure σ_v^M. After each structural finite element analysis, all the elements that satisfy the following condition are deleted:

$$\frac{\sigma_v^e}{\sigma_v^M} \leq RR_i, \tag{1.5}$$

where RR_i is the rejection ratio for iteration i. In each iteration i, the cycle of finite element analysis and element deletion is repeated using the same rejection ratio RR_i until a steady state is reached. After that, an evolutionary rate ER is added to the rejection ratio to obtain a new rejection ratio RR_{i+1} for iteration $i+1$,

$$RR_{i+1} = ER + RR_i. \tag{1.6}$$

Then, the cycle of finite element analysis and element deletion begins again, and the new rejection ratio is used until a new steady state is reached. The ESO method requires two parameters to be artificially prescribed: the initial rejection ratio and the evolutionary rate. For most designs, Xie et al. suggested using 1% for both parameters [112]. However, for some problems, much lower values must be used. In any case, after a few trials, it is not difficult to choose suitable values for these two parameters.

Common questions about the ESO method include whether it can ensure that a solution it obtains is not a local optimum and whether deleted elements can return. Zhou and Rozvany noted that the ESO method's rejection criteria may result in a highly suboptimal design [123]. To avoid this problem, Huang and Xie [37] suggested checking and maintaining the prescribed boundary conditions for the structure during each iteration of the optimization process. A new development for the ESO method is so-called bidirectional ESO (BESO), in which elements can be added as well as removed [67]. It has been also stated that the BESO method allows a search of all possible directions and thereby ensures a global optimum solution [108].

In the BESO method, it is necessary to identify the conditions under which an element must be added. This can also be done by comparing the von Mises stress of the element σ_v^e to the maximum von Mises stress of the structure σ_v^M. Elements can be added to the structure if they satisfy

$$\frac{\sigma_v^e}{\sigma_v^M} \geq IR_i, \tag{1.7}$$

where IR_i is the inclusion ratio [67].

The ESO method has been developed into a very effective engineering design tool and successfully applied to a variety of design problems, such as structures with stiffness or displacements constraints [112], frequency optimization [94] and micro-structures of cellular materials and composites [38]. An overview of research using the ESO or BESO methods for topology optimization can be found in [108].

However, one can find few papers that relate to using the ESO and BESO methods for topology optimization of compliant mechanisms. The primary study is [8]; Ansola et al. used ESO for compliant mechanism design. It demonstrated that the topologies of the obtained compliant mechanisms are similar to the topologies of those obtained using the SIMP method. Ansola et al. [6] proposed an element addition strategy for the topology optimization of thermally actuated compliant mechanisms. Recently, Alonso et al. [7] developed a topology optimization method involving a sequential element rejection and admission method for designing multi-material compliant mechanisms. Overall, ESO/BESO-based topology optimization of compliant mechanisms remains an open area that requires further investigation.

Level set methods The other major class of approaches to topology optimization uses structural boundaries as the design variables. Structural boundaries can be represented either using explicit boundary descriptions (e.g., spline-based representations) [97] or implicitly as iso-contours of a level set function [1, 96, 126]. In the context of spline-based methods, non-gradient-based methods (such as genetic algorithms) are often adopted to obtain the optimum configuration [97]. The disadvantage is that they are computationally inefficient.

Recently, level set methods have become powerful schemes for solving topology optimization problems [1, 15, 96, 126]. The level set method was originally introduced by Osher and Sethian [65]. Readers interested in applying level set methods are encouraged to read [63, 78]. The seminal works incorporating level set methods into structural optimization are [64, 79]. In a conventional level set method, optimization is implicitly accomplished by evolving a level set equation, i.e., a so-called Hamilton-Jacobi-type PDE (partial differential equation). The underlying idea can be expressed as follows.

Suppose that D is a design domain that completely contains the material domain Ω, and $D\backslash\Omega$ represents the void area. The idea underlying the level set method is to represent the structural boundaries Γ as the zero level set of a higher-dimensional function ϕ, as shown in Fig. 1.11. The reference (design) domain $D \subset R^d$ ($d = 2$ or 3), the material domain Ω, the void domain $D\backslash\Omega$ and the boundary Γ are

$$\begin{cases} \phi(\mathbf{x}, t) > 0 \text{ if } \mathbf{x} \in \Omega \\ \phi(\mathbf{x}, t) = 0 \text{ if } \mathbf{x} \in \Gamma \\ \phi(\mathbf{x}, t) < 0 \text{ if } \mathbf{x} \in D\backslash\Omega \end{cases} \tag{1.8}$$

where $\mathbf{x}$ is a point in the design domain and t is the pseudo-time [1, 15, 96].

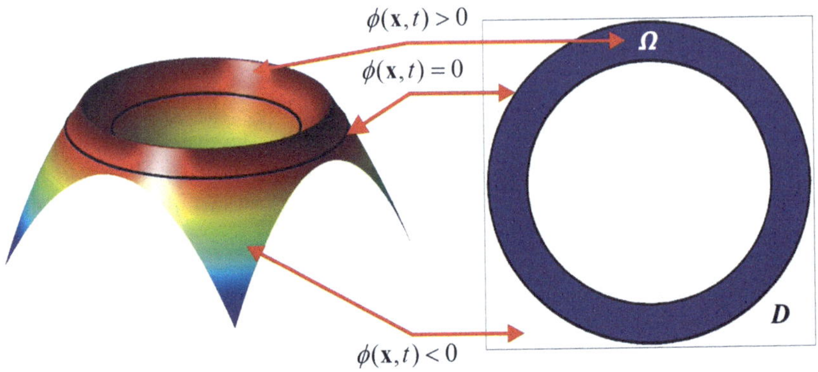

Fig. 1.11 The implicit level set function and its boundary

The process of level set-based optimization can be described by letting the level set function ϕ change dynamically over time. The motion of the interface is determined by a velocity field $\mathbf{V}$. Note that ϕ is a scalar function in R^3, which significantly reduces the complexity of topological changes, such as merging and pinching. Furthermore, the results of most level set-based topology optimization methods do not suffer from mesh dependency, which is often encountered in density-based topology optimization methods [13].

The key for moving the implicit surface is to find the velocity field $\mathbf{V}$, which, in our case, can be derived by means of a shape sensitivity analysis [18, 87]. Therefore, the evolution equation for the level set function ϕ is given by

$$\frac{\partial \phi}{\partial t} + \mathbf{V}\nabla\phi = 0. \tag{1.9}$$

The change in the embedded geometric shape is only related to the normal component of the velocity field V_n, whereas its tangential component V_t only influences its parametrization. With the unit normal $\mathbf{n} = \frac{\nabla\phi}{|\nabla\phi|}$, equation (1.9) can be rewritten as

$$\frac{\partial \phi}{\partial t} + V_n|\nabla\phi| = 0. \tag{1.10}$$

Equations (1.9) and (1.10) are the Hamilton-Jacobi-type partial differential equations known as the level set equations. In level set-based topology optimization, for example, the ENO2 (second-order essentially nonoscillatory scheme) [63] can be used to solve the Hamilton-Jacobi equation discretely.

Consider the problem of minimizing an objective functional J subject to a volume constraint functional G

$$\begin{aligned} \min: \quad & J = \int_D F(u) d\Omega \\ \text{s.t.}: \quad & G = Vol - Vol^* \leq 0. \end{aligned} \tag{1.11}$$

Incorporating the level set method, this optimization problem can be reformulated as follows:

$$\begin{aligned} \min: \quad & J(u,\phi) = \int_D F(u) H(\phi) d\Omega \\ \text{s.t.}: \quad & G(\phi) = \int_D H(\phi) d\Omega - Vol_{max} \leq 0, \end{aligned} \tag{1.12}$$

where u is the state variable and $H(\phi)$ is the Heaviside function,

$$H(\phi) = \begin{cases} 1 & \text{if} \quad \phi > 0 \\ 0 & \text{if} \quad \phi \leq 0 \end{cases} \tag{1.13}$$

To avoid numerical difficulties, the Heaviside function is approximated using equation [96]

$$H(\phi) = \begin{cases} \varepsilon & \text{if} \quad \phi < -\Lambda \\ (\frac{3}{4} - \frac{3\varepsilon}{4})(\frac{\phi}{\Lambda} - \frac{\phi^3}{3\Lambda^3}) + \frac{1+\varepsilon}{2} & \text{if} \quad -\Lambda \leq \phi \leq \Lambda \\ 1 & \text{if} \quad \phi > \Lambda \end{cases} \tag{1.14}$$

where ε is small to ensure that the numerical stiffness is nonsingular. Normally, it is set to 10^{-3}. Λ is set to $0.75\Delta x$.

To solve this optimization problem using the level set method, a common strategy is to apply the deepest gradient method. Define a Lagrangian function L and let λ be the Lagrange multiplier for the constraint. Then, the optimization problem can be reformulated as

$$L(u,\phi) = J(u,\phi) + \lambda\left(\int_D H(\phi) d\Omega - Vol_{max}\right). \tag{1.15}$$

The optimal result can be obtained by solving this optimization problem. The shape gradient of L is simply

$$\left\langle \frac{\partial L(u,\phi)}{\partial \phi}, \varphi \right\rangle = \left\langle \frac{\partial J(u,\phi)}{\partial \phi}, \varphi \right\rangle + \lambda \left\langle \frac{\partial \int_D H(\phi) d\Omega}{\partial \phi}, \varphi \right\rangle, \tag{1.16}$$

where the notation $\left\langle \frac{\partial L(u,\phi)}{\partial \phi}, \varphi \right\rangle$ represents the Fréchet derivative of the regularized Lagrangian L with respect to ϕ in the direction of φ. In fact, the shape derivative of L can be further expressed as

$$\left\langle \frac{\partial L(u,\phi)}{\partial \phi}, \varphi \right\rangle = \int_{\partial\Omega} l V_n ds, \tag{1.17}$$

where l is the shape gradient density [124]. To update the level set function, the velocity field is simply set to the negative of the shape gradient density, i.e.,

$$V_n = -l; \tag{1.18}$$

this choice guarantees that L decreases so that an optimum can be obtained.

Although the level set method has been accepted as a tool for solving topology optimization problems, it has several unfavorable features, such as the fact that new holes cannot be generated during optimization, the Courant-Friedrich-Levy (CFL) condition and the need for re-initialization. To make generating new holes possible, a topological derivative scheme has been integrated into the standard level set-based topology optimization method [33]. It is possible to update the level set function using topological gradients and to insert new holes by incorporating the ESO method [129].

Re-initialization is time-consuming because additional PDEs are solved in each iteration. Furthermore, because of the CFL condition, the step size for updating the level set function is determined by the smallest grid size, which is generally very small. Therefore, many iterations are often required to obtain an accurate optimal solution. An excessive number of iterations makes the optimization process computationally expensive. To improve the computational efficiency, a velocity mapping algorithm was developed by Mei and Wang [59] to speed up the evolution. Wang et al. [104] proposed two numerical techniques, velocity mapping and variational regularization, for improving the computational efficiency. However, more than 100 iterations are required to obtain an accurate design. Luo et al. [52] developed a semi-implicit level set method for structural topology optimization that allows enhanced relaxation based on the step size, which is initially limited by the CFL condition of the explicit scheme; therefore, the computational efficiency is significantly better. Recently, Zhou and Wang [124] proposed a semi-Lagrangian level set method for structural optimization. In their method, a line search algorithm and a sensitivity modulation scheme are introduced to require fewer design iterations to improve the overall computational efficiency. It is also possible to improve the computational efficiency by filtering the shape sensitivity [127]. Level set methods without re-initialization were developed by Zhu and Zhang [125, 128], who used energy functionals to maintain the level set function as a signed distance function (SDF); thus, re-initialization can be avoided.

Efforts also have made to develop more efficient strategies for level set-based topology optimization. Methods derived from image segmentation have been developed in [40, 53, 54, 103, 105]. In those methods, the level set function is defined as a piecewise continuous function throughout the design domain. This keeps the advantages of an implicit level set representation and avoids the need to solve the Hamilton-Jacobi equation.

Level set-based topology optimization has been applied to an increasing variety of design problems. For example, level set-based topology optimization of compliant mechanisms has involved thickness control [2, 31] and simultaneous consideration of support [109]. Optimization problems involving multiple materials can be easily solved using a level set approach [51]. Eigenvalue problems [114], contact problems

[60] and problems with design-dependent loads [113] have also been studied by using the level set method. In addition, the robustness of designs involving uncertainty has been investigated [16]. In addition to problems associated with compliance minimization and compliant mechanism design optimization, the level set method has been applied to physical problems of other types, such as thermal [58, 130, 132] and optical problems [29, 33]. For more information about level set methods for structural topology optimization, readers are referred to [22].

References

1. Allaire, G., Jouve, F.: Optimal design of micro-mechanisms by the homogenization method. Revue Européenne des Eléments **11**(2–4), 405–416 (2002)
2. Allaire, G., Jouve, F., Michailidis, G.: Thickness control in structural optimization via a level set method. Structural Multidisciplinary Optimization **53**(6), 1349–1382 (2016)
3. Allaire, G., Jouve, F., Toader, A.M.: A level-set method for shape optimization. C.R. Math. **334**(12), 1125–1130 (2002)
4. Ananthasuresh, G., Kota, S., Gianchandani, Y.: Systematic synthesis of microcompliant mechanisms-preliminary results. In: National Applied Mechanisms and Robotics Conference, Cincinnati, OH, November, Paper, vol. 82 (1993)
5. Andersen, K.N., Carlson, K., Petersen, D.H., Eichhorn, V., Fatikow, S.: Electrothermal micro-grippers for pick-and-place operations. Microelectron. Eng. **85**(5–6), 1128–1130 (2008)
6. Ansola, R., Vegueria, E., Canales, J.: An element addition strategy for thermally actuated compliant mechanism topology optimization. Eng. Comput. **27**(6), 694–711 (2010)
7. Ansola, R., Veguería, E., Canales, J., Alonso, C.: Evolutionary optimization of compliant mechanisms subjected to non-uniform thermal effects. Finite Elem. Anal. Des. **57**, 1–14 (2012)
8. Ansola, R., Veguería, E., Canales, J., Tárrago, J.A.: A simple evolutionary topology optimization procedure for compliant mechanism design. Finite Elem. Anal. Des. **44**(1), 53–62 (2007)
9. Azamirad, G., Arezoo, B.: Structural design of stamping die components using bi-directional evolutionary structural optimization method. Int. J. Adv. Manufacturing Technol. **87**(1–4), 969–979 (2016)
10. Barthelemy, J.F., Haftka, R.T.: Approximation concepts for optimum structural designła review. Structural Optimization **5**(3), 129–144 (1993)
11. Bendsøe, M.P., Kikuchi, N.: Generating optimal topologies in structural design using a homogenization method. Comput. Methods Appl. Mech. Eng. **71**(2), 197–224 (1988)
12. Bendsøe, M.P., Sigmund, O.: Material interpolation schemes in topology optimization. Arch. Appl. Mech. **69**(9), 635–654 (1999)
13. Bendsøe, M.P., Sigmund, O.: Topology Optimization: Theory, Methods and Applications. Springer, Heidelberg (2004)
14. Bogue, R.: Recent developments in mems sensors: a review of applications, markets and technologies. Sensor Rev. **33**(4), 300–304 (2013)
15. Challis, V.J.: A discrete level-set topology optimization code written in Matlab. Structural Multidisciplinary Optimization **41**(3), 453–464 (2010)
16. Chen, S.K., Gonella, S., Chen, W., Liu, W.K.: A level set approach for optimal design of smart energy harvesters. Comput. Methods Appl. Mech. Eng. **199**(37–40), 2532–2543 (2010)
17. Cho, S., Kwak, J.: Topology design optimization of geometrically non-linear structures using meshfree method. Comput. Methods Appl. Mech. Eng. **195**(44), 5909–5925 (2006)
18. Choi, K.K., Kim, N.H.: Structural Sensitivity Analysis and Optimization 1: Linear Systems. Springer Science & Business Media (2006)

19. Choi, Y., Ross, J., Wester, B., Allen, M.G.: Mechanically driven microtweezers with integrated microelectrodes. J. Micromech. Microeng. **18**(6), 065,004 (2008)
20. Deaton, J.D., Grandhi, R.V.: A survey of structural and multidisciplinary continuum topology optimization: post 2000. Structural Multidisciplinary Optimization **49**(1), 1–38 (2014)
21. Diaz, A., Sigmund, O.: Checkerboard patterns in layout optimization. Structural Optimization **10**(1), 40–45 (1995)
22. van Dijk, N.P., Maute, K., Langelaar, M., Van Keulen, F.: Level-set methods for structural topology optimization: a review. Structural Multidisciplinary Optimization **48**(3), 437–472 (2013)
23. Dorn, W.S.: Automatic design of optimal structures. Journal de Mecanique **3**, 25–52 (1964)
24. Eschenauer, H.A., Kobelev, V.V., Schumacher, A.: Bubble method for topology and shape optimization of structures. Structural Optimization **8**(1), 42–51 (1994)
25. Fancello, E.A.: Topology optimization for minimum mass design considering local failure constraints and contact boundary conditions. Structural Multidisciplinary Optimization **32**(3), 229–240 (2006)
26. Fatikow, S., Rembold, U.: Microsystem Technology and Microrobotics. Springer (1997)
27. Fleury, C.: Structural weight optimization by dual methods of convex programming. Int. J. Numer. Meth. Eng. **14**(12), 1761–1783 (1979)
28. Frecker, M., Ananthasuresh, G., Nishiwaki, S., Kikuchi, N., Kota, S.: Topological synthesis of compliant mechanisms using multi-criteria optimization. J. Mech. Des. **119**(2), 238–245 (1997)
29. Frei, W., Tortorelli, D., Johnson, H.: Geometry projection method for optimizing photonic nanostructures. Opt. Lett. **32**(1), 77–79 (2007)
30. Furukawa, E., Mizuno, M.: Piezo-driven translation mechanisms utilizing linkages. J. Jpn. Soc. Precis. Eng. **58**(1), 127–132 (1992)
31. Guo, X., Zhang, W., Zhong, W.: Explicit feature control in structural topology optimization via level set method. Comput. Methods Appl. Mech. Eng. **272**, 354–378 (2014)
32. Hassini, B., Hinton, E.: A review of homogenization and topology optimization I to III. Comput. Struct. **69**, 707–756 (1998)
33. He, L., Kao, C.Y., Osher, S.: Incorporating topological derivatives into shape derivatives based level set methods. J. Comput. Phys. **225**(1), 891–909 (2007)
34. Heo, S., Kim, Y.Y.: Optimal design and fabrication of mems rotary thermal actuators. J. Micromech. Microeng. **17**(11), 2241 (2007)
35. Howell, L.L.: Compliant Mechanisms. Wiley (2001)
36. Howell, L.L., Midha, A.: Parametric deflection approximations for end-loaded, large-deflection beams in compliant mechanisms. ASME J. Mech. Des. **117**(1), 156–165 (1995)
37. Huang, X., Xie, Y.M.: A further review of ESO type methods for topology optimization. Structural Multidisciplinary Optimization **41**(5), 671–683 (2010)
38. Huang, X., Zhou, S., Xie, Y., Li, Q.: Topology optimization of microstructures of cellular materials and composites for macrostructures. Comput. Mater. Sci. **67**, 397–407 (2013)
39. Jensen, B.D., Howell, L.L.: The modeling of cross-axis flexural pivots. Mech. Mach. Theory **37**(5), 461–476 (2002)
40. Jiang, L., Chen, S.: Parametric structural shape & topology optimization with a variational distance-regularized level set method. Comput. Methods Appl. Mech. Eng. **321**, 316–336 (2017)
41. Jog, C.S., Haber, R.B.: Stability of finite element models for distributed-parameter optimization and topology design. Comput. Methods Appl. Mech. Eng. **130**(3–4), 203–226 (1996)
42. Joo, J., Kota, S.: Topological synthesis of compliant mechanisms using nonlinear beam elements. Mech. Based Des. Structures Mach. **32**(1), 17–38 (2004)
43. Joo, J., Kota, S., Kikuchi, N.: Topological synthesis of compliant mechanisms using linear beam elements. Mech. Structures Mach. **28**(4), 245–280 (2000)
44. Kinoshita, T., Ohsaki, M.: Synthesis of bistable compliant structures from Truss mechanisms. J. Comput. Sci. Technol. **3**(2), 417–425 (2009)

45. Kota, S., Ananthasuresh, G.: Designing compliant mechanisms. Mech. Eng. CIME **117**(11), 93–97 (1995)
46. Lee, K.M., Arjunan, S.: A three-degrees-of-freedom micromotion in-parallel actuated manipulator. IEEE Trans. Robot. Automation **7**(5), 634–641 (1991)
47. Li, H., Zhang, X., Wu, H., Gan, J.: Line-based calibration of a micro-vision motion measurement system. Opt. Lasers Eng. **93**, 40–46 (2017)
48. Li, H., Zhang, X.M., Zeng, L., Huang, Y.J.: A monocular vision system for online pose measurement of a 3RRR planar parallel manipulator. J. Intell. Rob. Syst. **2**, 1–15 (2017)
49. Li, Z., Zhang, X.: Reliability-based topology optimization of compliant micro-gripper with geometrical nonlinearity. J South China Univ. Technol. (Natural Science Edition) **8**, 023 (2008)
50. Li, Z., Zhang, X.: Topology optimization of multiple inputs and outputs compliant mechanisms with geometrically nonlinearity. Chin. J. Mech. Eng. **45**(1), 180–188 (2009)
51. Liu, P., Luo, Y., Kang, Z.: Multi-material topology optimization considering interface behavior via XFEM and level set method. Comput. Methods Appl. Mech. Eng. **308**, 113–133 (2016)
52. Luo, J., Luo, Z., Chen, L., Tong, L., Wang, M.Y.: A semi-implicit level set method for structural shape and topology optimization. J. Comput. Phys. **227**(11), 5561–5581 (2008)
53. Luo, Z., Tong, L., Wang, M.Y., Wang, S.: Shape and topology optimization of compliant mechanisms using a parameterization level set method. J. Comput. Phys. **227**(1), 680–705 (2007)
54. Luo, Z., Wang, M.Y., Wang, S., Wei, P.: A level set-based parameterization method for structural shape and topology optimization. Int. J. Numer. Meth. Eng. **76**(1), 1–26 (2008)
55. Ma, Z.D., Kikuchi, N., Cheng, H.C.: Topological design for vibrating structures. Comput. Methods Appl. Mech. Eng. **121**(s 1C4), 259–280 (1995)
56. Ma, Z.D., Kikuchi, N., Hagiwara, I.: Structural topology and shape optimization for a frequency response problem. Comput. Mech. **13**(3), 157–174 (1993)
57. Mankame, N.D., Ananthasuresh, G.: Topology synthesis of electrothermal compliant mechanisms using line elements. Structural Multidisciplinary Optimization **26**(3), 209–218 (2004)
58. Maute, K., Kreissl, S., Makhija, D., Yang, R.: Topology optimization of heat conduction in nano-composites. In: 9th World Congress on Structural and Multidisciplinary Optimization, Shizuoka, Japan (2011)
59. Mei, Y., Wang, X.: A level set method for structural topology optimization and its applications. Adv. Eng. Softw. **35**(7), 415–441 (2004)
60. Myśliński, A.: Level set method for optimization of contact problems. Eng. Anal. Boundary Elem. **32**(11), 986–994 (2008)
61. Nishiwaki, S., Frecker, M.I., Min, S., Kikuchi, N.: Topology optimization of compliant mechanisms using the homogenization method. Int. J. Numer. Meth. Engng. **42**, 535–559 (1998)
62. Novotny, A.A., Sokołowski, J.: Topological Derivatives in Shape Optimization. Springer Science & Business Media (2012)
63. Osher, S., Fedkiw, R.: Level Set Methods and Dynamic Implicit Surfaces. Springer, New York (2002)
64. Osher, S., Santosa, F.: Level-set methods for optimization problem involving geometry and constraints: I. Frequencies of a two-density inhomogeneous drum. J. Comput. Phys. **171**(1), 272–288 (2001)
65. Osher, S., Sethian, J.A.: Fronts propagating with curvature-dependent speed: algorithms based on Hamilton-Jacobi formulations. J. Comput. Phys. **79**(1), 12–49 (1988)
66. Ouyang, P.R., Tjiptoprodjo, R.C., Zhang, W.J., Yang, G.S.: Micro-motion devices technology: the state of arts review. Int. J. Adv. Manuf. Technol. **38**(5–6), 463–478 (2008)
67. Querin, O., Steven, G., Xie, Y.: Evolutionary structural optimisation (ESO) using a bidirectional algorithm. Eng. Comput. **15**(8), 1031–1048 (1998)
68. Rai, A.K., Saxena, A., Mankame, N.D.: Unified synthesis of compact planar path-generating linkages with rigid and deformable members. Structural Multidisciplinary Optimization **41**(6), 863 (2010)

69. Ramrakhyani, D.S., Frecker, M.I., Lesieutre, G.A.: Hinged beam elements for the topology design of compliant mechanisms using the ground structure approach. Structural Multidisciplinary Optimization **37**(6), 557–567 (2009)
70. Rietz, A.: Sufficiency of a finite exponent in simp (power law) methods. Structural Multidisciplinary Optimization **21**(2), 159–163 (2001)
71. Rozvany, G.: Stress ratio and compliance based methods in topology optimization-a critical review. Structural Multidisciplinary Optimization **21**(2), 109–119 (2001)
72. Rozvany, G.I.: A critical review of established methods of structural topology optimization. Structural Multidisciplinary Optimization **37**(3), 217–237 (2009)
73. Sardan, O., Eichhorn, V., Petersen, D., Fatikow, S., Sigmund, O., Bøggild, P.: Rapid prototyping of nanotube-based devices using topology-optimized microgrippers. Nanotechnology **19**(49), 495,503 (2008)
74. Saxena, A., Ananthasuresh, G.: On an optimal property of compliant topologies. Structural Multidisciplinary Optimization **19**(1), 36–49 (2000)
75. Saxena, A., Ananthasuresh, G.: Topology optimization of compliant mechanisms with strength considerations. Mech. Structures Mach. **29**(2), 199–221 (2001)
76. Saxena, A., Ananthasuresh, G.: Topology synthesis of compliant mechanisms for nonlinear force-deflection and curved path specifications. J. Mech. Des. **123**(1), 33–42 (2001)
77. Scire, F.E., Teague, E.C.: Piezodriven 50-μm range stage with subnanometer resolution. Rev. Sci. Instrum. **49**(12), 1735 (1978)
78. Sethain, J.A.: Level Set Methods and Fast Marching Methods: Evolving Interfaces in Computational Geometry, Fluid Mechanics, Omputer Version, and Material Science. Cambridge University Press, Cambridge (1999)
79. Sethian, J.A., Wiegmann, A.: Structural boundary design via level set and immersed interface methods. J. Comput. Phys. **163**(2), 489–528 (2000)
80. Shi, H., Shi, W., Zhang, R., Zhai, J., Chu, J., Dong, S.: A micromachined piezoelectric microgripper for manipulation of micro/nanomaterials. Rev. Sci. Instrum. **88**(6), 235 (2017)
81. Sigmund, O.: Design of Materials Structures Using Topology Optimization. Technical University of Denmark, Department of Solid Mechanics (1994)
82. Sigmund, O.: On the design of compliant mechanisms using topology optimization. Mech. Structures Mach. **25**(4), 493–524 (1997)
83. Sigmund, O.: Design of multiphysics actuators using topology optimization-part I: One-material structures. Comput. Methods Appl. Mech. Eng. **190**(49), 6577–6604 (2001)
84. Sigmund, O.: Morphology-based black and white filters for topology optimization. Structural Multidisciplinary Optimization **33**(4–5), 401–424 (2007)
85. Sigmund, O.: On the usefulness of non-gradient approaches in topology optimization. Structural Multidisciplinary Optimization **43**(5), 589–596 (2011)
86. Sigmund, O., Petersson, J.: Numerical instabilities in topology optimization: a survey on procedures dealing with checkerboards, mesh-dependencies and local minima. Structural Optimization **16**(1), 68–75 (1998)
87. Sokolowski, J., Zolesio, J.P.: Introduction to Shape Optimization. In: Introduction to Shape Optimization, pp. 5–12. Springer (1992)
88. Svanberg, K.: The method of moving asymptotesła new method for structural optimization. Int. J. Numer. Meth. Eng. **24**(2), 359–373 (1987)
89. Takezawa, A., Nishiwaki, S., Kitamura, M.: Shape and topology optimization based on the phase field method and sensitivity analysis. J. Comput. Phys. **229**(7), 2697–2718 (2010)
90. Tang, H., Li, Y.: Development and active disturbance rejection control of a compliant micro-/nanopositioning piezostage with dual mode. IEEE Trans. Industr. Electron. **61**(3), 1475–1492 (2013)
91. Tanskanen, P.: The evolutionary structural optimization method: theoretical aspects. Comput. Methods Appl. Mech. Eng. **191**(47), 5485–5498 (2002)
92. Tran, A.V., Zhang, X., Zhu, B.: The development of a new piezoresistive pressure sensor for low pressures. IEEE Tran. Industrial Electronics (2017)

93. Uicker, J.J., Pennock, G.R., Shigley, J.E., et al.: Theory of Machines and Mechanisms, vol. 1. Oxford University Press, New York (2011)
94. Vicente, W.M., Picelli, R., Pavanello, R., Xie, Y.M.: Topology optimization of frequency responses of fluidcstructure interaction systems. Finite Elem. Anal. Des. **98**, 1–13 (2015)
95. Wang, H., Zhang, X.: Input coupling analysis and optimal design of a 3-DOF compliant micro-positioning stage. Mech. Mach. Theory **43**(4), 400–410 (2008)
96. Wang, M.Y., Wang, X., Guo, D.: A level set method for structural topology optimization. Comput. Methods Appl. Mech. Eng. **192**(1), 227–246 (2003)
97. Wang, N., Zhang, X.: Compliant mechanisms design based on pairs of curves. Sci. China Technol. Sci. **55**(8), 2099–2106 (2012)
98. Wang, R., Zhang, X.: Design and test of a novel planar 3-DOF precision positioning platform with a large magnification. In: International Conference on Manipulation, Manufacturing and Measurement on the Nanoscale, pp. 236–243 (2014)
99. Wang, R., Zhang, X.: Preload characteristics identification of the piezoelectric- actuated 1-DOF compliant nanopositioning platform. Frontiers Mech. Eng. **10**(1), 20–36 (2015)
100. Wang, R., Zhang, X.: A planar 3-DOF nanopositioning platform with large magnification. Precis. Eng. **46**, 221–231 (2016)
101. Wang, R., Zhang, X.: Optimal design of a planar parallel 3-DOF nanopositioner with multi-objective. Mech. Mach. Theory **112**, 61–83 (2017)
102. Wang, R., Zhang, X.: Parameters optimization and experiment of a planar parallel 3-dof nanopositioning system. IEEE Trans. Industrial Electron. **PP**(99), 1–1 (2017)
103. Wang, S., Wang, M.Y.: Radial basis functions and level set method for structural topology optimization. Int. J. Numer. Meth. Eng. **65**(12), 2060–2090 (2006)
104. Wang, X., Wang, M., Guo, D.: Structural shape and topology optimization in a level-set-based framework of region representation. Structural Multidisciplinary Optimization **27**(1–2), 1–19 (2004)
105. Wei, P., Wang, M.Y.: Piecewise constant level set method for structural topology optimization. Int. J. Numer. Meth. Eng. **78**(4), 379–402 (2009)
106. Wittwer, J.W., Chase, K.W., Howell, L.L.: The direct linearization method applied to position error in kinematic linkages. Mech. Mach. Theory **39**(7), 681–693 (2004)
107. Wu, T.L., Chen, J.H., Chang, S.H.: A six-dof prismatic-spherical-spherical parallel compliant nanopositioner. IEEE Trans. Ultrason. Ferroelectr. Freq. Control **55**(12), 2544 (2008)
108. Xia, L., Xia, Q., Huang, X., Xie, Y.M.: Bi-directional evolutionary structural optimization on advanced structures and materials: a comprehensive review. Arch. Comput. Methods Eng. 1–42 (2016)
109. Xia, Q., Shi, T.: Topology optimization of compliant mechanism and its support through a level set method. Comput. Methods Appl. Mech. Eng. **305**, 359–375 (2016)
110. Xie, Y., Steven, G.: Evolutionary structural optimization for dynamic problems. Comput. Structures **58**(6), 1067–1073 (1996)
111. Xie, Y.M., Steven, G.P.: A simple evolutionary procedure for structural optimization. Comput. Structures **49**(5), 885–896 (1993)
112. Xie, Y.M., Steven, G.P.: Basic evolutionary structural optimization. In: Evolutionary Structural Optimization, pp. 12–29. Springer, London (1997)
113. Yamada, T., Izui, K., Nishiwaki, S.: A level set-based topology optimization method for maximizing thermal diffusivity in problems including design-dependent effects. J. Mech. Des. **133**(3), 031,011 (2011)
114. Yamada, T., Izui, K., Nishiwaki, S., et al.: A topology optimization method based on the level set method incorporating a fictitious interface energy. Comput. Methods Appl. Mech. Eng. **199**(45), 2876–2891 (2010)
115. Yao, Q., Dong, J., Ferreira, P.M.: Design, analysis, fabrication and testing of a parallel-kinematic micropositioning XY stage. Int. J. Mach. Tools Manuf. **47**(6), 946–961 (2007)
116. Yu, Y.Q., Zhu, S.K.: 5R pseudo-rigid-body model for inflection beams in compliant mechanisms. Mech. Mach. Theory **116**, 501–512 (2017)

117. Zhan, J., Zhang, X.: Topology optimization of compliant mechanisms with geometrical nonlinearities using the ground structure approach. Chin. J. Mech. Eng. **24**(2), 1 (2011)
118. Zhang, X., Hu, K., Wang, N.e.a.: Multi-objective topology optimization of multiple materials compliant mechanisms based on parallel strategy. J. Mech. Eng. (in Chinese) **52**, 1–8 (2016)
119. Zhang, X., Ouyang, G., Wang, H.: Topology optimization of multiple inputs and multiple outputs compliant mechanisms. Chin. J. Mech. Eng. **20**(1), 82–85 (2007)
120. Zhang, X., Zheng, Y., Ota, J., Huang, Y.: Peg-in-hole assembly based on two-phase scheme and F/T sensor for dual-arm robot. Sensors **17**(9), 2004 (2017)
121. Zhou, M., Rozvany, G.: Dcoc: an optimality criteria method for large systems part I: theory. Structural Optimization **5**(1–2), 12–25 (1992)
122. Zhou, M., Rozvany, G.: Dcoc: an optimality criteria method for large systems part II: algorithm. Structural Optimization **6**(4), 250–262 (1993)
123. Zhou, M., Rozvany, G.: On the validity of ESO type methods in topology optimization. Structural Multidisciplinary Optimization **21**(1), 80–83 (2001)
124. Zhou, M., Wang, M.Y.: A semi-lagrangian level set method for structural optimization. Structural Multidisciplinary Optimization **46**(4), 487–501 (2012)
125. Zhu, B., Zhang, X.: A new level set method for topology optimization of distributed compliant mechanisms. Int. J. Numer. Meth. Eng. **91**(8), 843–871 (2012)
126. Zhu, B., Zhang, X., Fatikow, S.: Level set-based topology optimization of hinge-free compliant mechanisms using a two-step elastic modeling methodhanisms using a two-step elastic modeling method. ASME J. Mech. Des. **136**(2), 031,007 (2014)
127. Zhu, B., Zhang, X., Fatikow, S.: Filter the shape sensitivity in level set-based topology optimization methods. Structural Multidisciplinary Optimization **51**(5), 1035–1049 (2015)
128. Zhu, B., Zhang, X., Fatikow, S.: Structural topology and shape optimization using a level set method with distance-suppression scheme. Comput. Methods Appl. Mech. Eng. **283**, 1214–1239 (2015)
129. Zhu, B., Zhang, X., Fatikow, S., Wang, N.: Bi-directional evolutionary level set method for topology optimization. Engineering Optimization **47**(3), 390–406 (2015)
130. Zhu, B., Zhang, X., Wang, N., Fatikow, S.: Optimize heat conduction problem using level set method with a weighting based velocity constructing scheme. Int. J. Heat Mass Transf. **99**, 441–451 (2016)
131. Zhu, S.K., Yu, Y.Q.: Pseudo-rigid-body model for the flexural beam with an inflection point in compliant mechanisms. J. Mech. Robot. (2017)
132. Zhuang, C., Xiong, Z., Ding, H.: A level set method for topology optimization of heat conduction problem under multiple load cases. Comput. Methods Appl. Mech. Eng. **196**(4), 1074–1084 (2007)

Chapter 2
Topology Optimization of Flexure Hinges

Abstract This chapter is devoted to developing a systematic method for the topology optimization of flexure hinges. The main research work includes: (1) Developing basic topology optimization models for translational and revolute flexure hinges. The objective function for the topology optimization of flexure hinges is to maximize the compliance in the rotational direction while minimizing the compliance in the direction of parasitic motion. (2) Reducing the maximum stress and stress concentration factor by considering the stress constraint in the optimization model. (3) Improving the performances of the common notch flexure hinges by redesigning their configurations using topology optimization approach.

2.1 Introduction

A large number of compliant mechanisms are constructed based on flexure hinges, especially those synthesized by using the kinematic-based approach. The flexure hinge, alternatively called the flexural pivot, is a thin member that provides relative rotation between two adjacent rigid members through flexing (bending). Flexure hinges have several advantages over conventional rotational joints because they are monolithic with the rest of the mechanism [12, 22]. Additionally, they are compact and have no friction losses, no need for lubrication, and no backlash; therefore, flexure hinges are widely used in micropositioning stages [40], high-accuracy alignment devices for optical fibers [6], etc. Several flexure hinges with different configurations have been developed over the past decades, and some of them are shown in Fig. 2.1.

Since Paros and Weisbord [26] developed a thorough analytical presentation of the conventional right-circular flexure hinge, substantial research has been focused on deriving the theoretical equations of flexure hinges with other shapes. The design of new types of compliant hinges with a large displacement [37] is also an important research direction because of the limitations in the range of motion of conventional flexure hinges.

In particular, Smith et al. [32] presented the compliance equation for the elliptical flexure hinge and checked the model prediction using finite element analysis and experimental data. Stress concentration factors for the elliptical hinge were also

X. Zhang and B. Zhu, *Topology Optimization of Compliant Mechanisms*,
https://doi.org/10.1007/978-981-13-0432-3_2

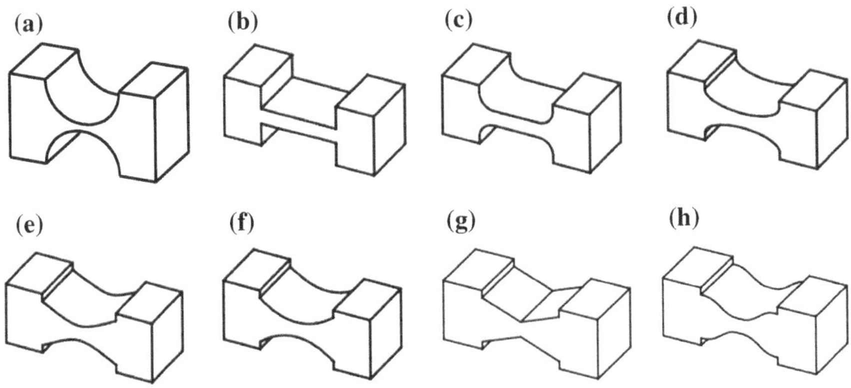

Fig. 2.1 Notch flexure hinges: **a** right-circular, **b** leaf, **c** corner-filleted, **d** elliptic, **e** hyperbolic, **f** parabolic, **g** V-shaped, and **h** cycloidal

presented based on finite element analysis results. Lobontiu et al. [23] presented the closed-form solutions of compliance of the corner-filleted hinge based on Castigliano's first theorem. Lobontiu et al. developed closed-form compliance equations for conic-section (circular, elliptic, parabolic, and hyperbolic [24, 25]) flexure hinges. Tseytlin [38] presented effective compliance equations of monolithic flexure hinges with different cross sections based on conformal mapping theory. The accuracy was therefore improved compared with other theoretical equations. Awtar et al. [1] presented a general dimensionless analytical framework of beam-based flexure mechanisms and applied it to analyze several beam-based flexure modules.

A more reliable direct finite element analysis method was developed by Smith [31]. Based on finite element analysis, Schotborgh et al. [29] presented dimensionless design graphs for circular, cross and beam hinges to determine the optimal geometry of the hinges. Bona and Munteanu [8] developed a method for the optimal design of flexure hinges by coupling a parametric finite element model. The method was used for optimization of flexure hinges at the shape level, that is, the topology of the hinges could not be optimized. Later, Zelenika [45] presented a method for nonlinear parametric optimization of flexural hinge shapes based on finite analysis. Lin et al. [16] proposed a method for designing hybrid flexure hinges. A new flexure hinge composed of half of a hyperbolic flexure hinge and half of a corner-filleted flexure hinge was developed. The corresponding closed-form equations were formulated to characterize both the active rotation and all other in-plane parasitic motion by Castigliano's second theorem. The hybrid flexure hinges were quantitatively compared with five common types of notch flexure hinges (circular, corner-filleted, elliptical, hyperbolic, and parabolic) based on compliance, precision, compliance-precision ratio, and maximum stress.

The aforementioned design and optimization procedures were developed based on either a known topology (e.g., in Ref. [45]) or a known shape (e.g., in Ref. [23]). Therefore, the design procedures depend, to a large extent, on the designer's experience and inspiration. Screw theory-based methods have been developed to conceptually design flexible joints [11, 33, 44]; however, rigorous but abstruse mathematical knowledge is required, which makes the methods difficult to extend.

This part of the work is devoted to presenting an alternative method for the design of flexure hinges at the topology level using a topology optimization method. This chapter will introduce the flexure hinge optimization model, objective functions, constraint functions and the redesign of notch flexure hinges.

2.2 Optimization Problem Setting

Once the topology is selected, the shape (geometric profile) of the hinge becomes the key factor in determining the performance of the flexure hinge. Two types of flexure hinges are considered in this section: translational and revolute. Translational and revolute hinges are two types of flexure hinge that are widely used in compliant mechanisms to provide translational and rotational DOFs. Translational hinges have been designed as two primitive types: leaf spring (Fig. 2.2a) and notch (Fig. 2.2b). Both types are analogous to a four-bar linkage. The topology of two well-studied and widely used revolute hinges with specific geometric profiles are shown in Fig. 2.3.

Flexure hinges are generally modeled and analyzed as small-displacement beams. However, even with this simplified problem, two basic questions need to be answered. (1) Is the pre-set topology of the flexure hinge the best? (2) How can the best topology of the flexure hinge be determined, that is, how can a flexure hinge be optimized at the topology level?

The aim of this paper is to focus on the topology optimization of flexure hinges. Two types of flexure hinge are considered: the translational hinge and the revolute hinge. The following work is proposed based on several assumptions.

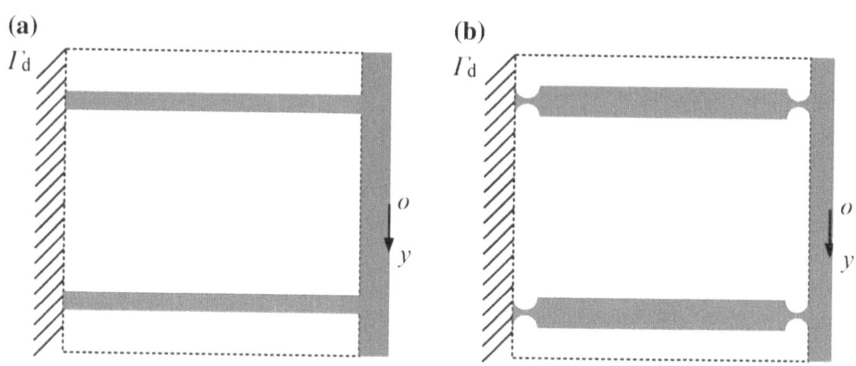

Fig. 2.2 Basic translational hinges with the same topology. **a** Leaf spring type and **b** notch type. In the figure, Γ_d is the Dirichlet (or displacement) boundary condition. When the hinge is loaded at point o with a force along the y-direction, displacement is produced along the y-direction [12]

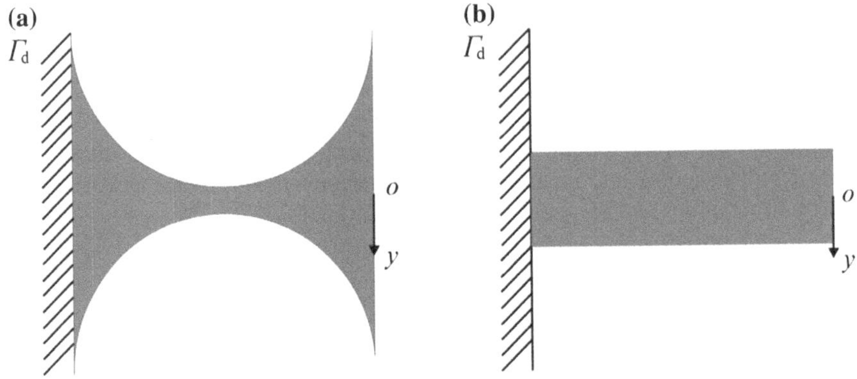

Fig. 2.3 Basic revolute hinges with the same topology. **a** Circular flexure hinge and **b** constant rectangular cross section flexure hinge. In the figure, Γ_d is the Dirichlet (or displacement) boundary condition. When the hinge is loaded at point o with a force along the y-direction, an angular displacement is produced [24, 25]

(1) The flexure hinges are designed to be used in two-dimensional monolithic compliant mechanisms.
(2) Optimization models are formulated to characterize only in-plane motion.
(3) The flexure hinges are subjected to bending by force, and shearing effects are not considered. Axis loading is considered to generate a well-posed design.
(4) The design domains of the two proposed types of flexure hinge are symmetric with respect to the longitudinal axis. The design domains are considered to be fixed at one end and free at the other.

2.2.1 *Optimization Model*

This section is devoted to developing the basic optimization models for the topology design of translational and revolute hinges [18, 46]. A general design domain for the topology optimization of flexure hinges is illustrated in Fig. 2.4, where y denotes the desired direction and x denotes the undesired direction. **D** is the design domain, and **R** is the rigid domain. The design domain **D** is fixed at the boundary Γ_d. An ideal flexure hinge should be as compliant as possible in the direction DOFs while being as stiff as possible in the other directions [44, 46]. Therefore, u_d due to F_d needs to be maximized while minimizing u_{ud} due to F_{ud}. An artificial spring k_s is added to avoid numerical instabilities. Additionally, other criteria, such as axis drift, must be established for specific applications [22].

The accuracy of the finite element analysis model is significantly influenced by how the boundary conditions are assigned on the design domain [43]. Hence, the width l of the rigid area **R** is set to be at least equal to or greater than the width L

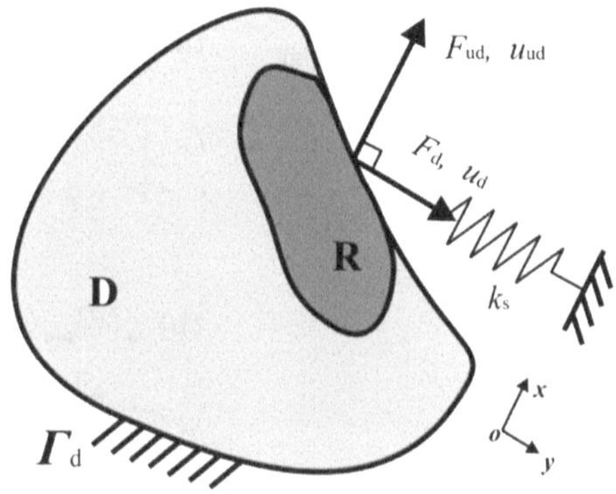

Fig. 2.4 The flexure hinge design domain

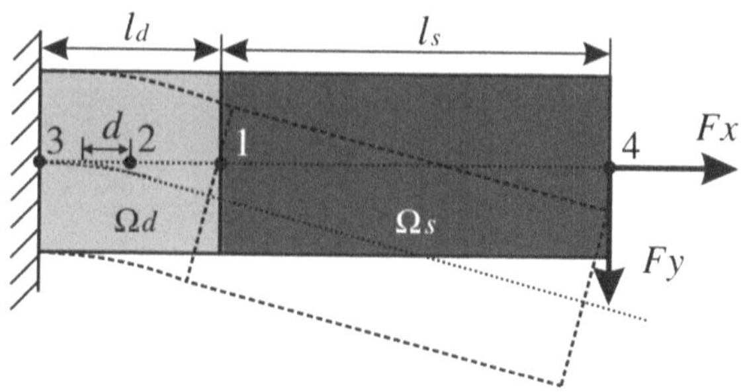

Fig. 2.5 The design domain of the flexure hinge considered in this chapter

of the design domain **D**. Forces are applied at a point where $l_s \geq 3l_d$, as is shown in Fig. 2.5.

2.2.2 *Translational Flexure Hinges*

When designing translational hinges, the goal is to find a constraint arrangement that provides approximate straight-line motion. One popular hinge that can accomplish this function is the parallelogram flexure hinge [1]. Here, we attempt to develop a general topology optimization model the translational hinge. For simplicity, F_x and F_y are set unity. As shown in Fig. 2.6a, when designing a translational hinge, the compliance C_y due to load F_y needs to be maximized and the compliance C_x due to load F_x needs to be minimized. C_x and C_y can be stated as follows:

$$C_x = \int_D E_{ijkl}\varepsilon_{ij}(u_x)\varepsilon_{kl}(u_x)d\Omega, \tag{2.1}$$

$$C_y = \int_D E_{ijkl}\varepsilon_{ij}(u_y)\varepsilon_{kl}(u_y)d\Omega, \tag{2.2}$$

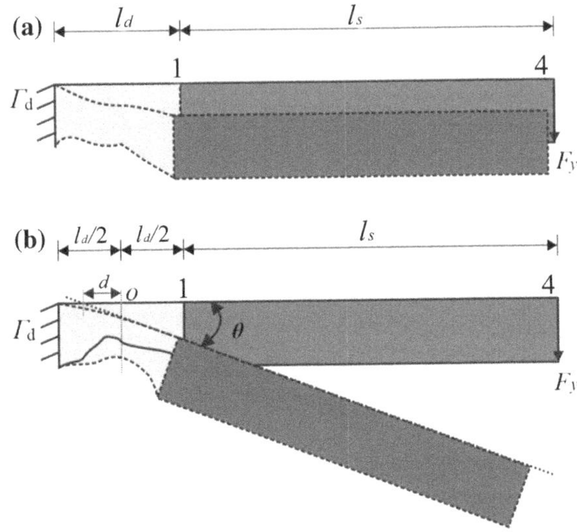

Fig. 2.6 Fictitious deformations of translational (**a**) and the revolute (**b**) hinges under F_y

where u_x and u_y are the displacement fields due to F_x and F_y, respectively. E_{ijkl} and ε are the stiffness and linearized strain tensors, respectively.

To satisfy the translational requirement, when a force is applied at point 4 along the y-direction, the rigid part **R** should move along the y-direction, as shown in Fig. 2.6a. Therefore, the constraint on the translational requirement can be written as

$$\eta^T = \frac{1}{2}\left(u_{F_y}^{4_y} - u_{F_y}^{1_y}\right)^2 \leq \eta_{obj}, \tag{2.3}$$

where $u_{F_y}^{4_y}$ denotes the displacement of point 4 along the y-direction due to load F_y, and $u_{F_y}^{1_y}$ denotes the displacement of point 1 along the y-direction due to load F_y. $\eta_{obj} \geq 0$ is a small constant.

In conclusion, the optimization model Q_T for topology optimization translational hinge can be defined as follows:

$$\begin{aligned} Q_T : \min : &\ -C_y + C_x \\ \text{s.t.} : &\ Vol \leq V^* \\ &\ \eta \leq \eta_{obj}. \end{aligned} \tag{2.4}$$

2.2.3 Revolute Flexure Hinges

The main requirement in the design of a revolute hinge is to fulfill finite rotation around the z-axis (perpendicular to the design domain).

The design condition is shown in Fig. 2.6b. The desired rotation along the z-axis should be maximized. The compliance C_x due to load F_x (undesired) should be minimized. Similarly to the design of the translational hinge, the compliance C_y due to load F_y needs to be maximized. Therefore, a possible objective function for the topology design of revolute hinges can be stated as

$$\min : \quad -C_y + C_x. \tag{2.5}$$

Figure 2.6(b) shows a fictitious deformation of the design domain due to F_y. As the horizontal axis is a symmetry axis, only half of the design domain is taken into consideration.

Generally, any hinges that make $u_{F_y}^{4_y}$ not equal to $u_{F_y}^{1_y}$ ($u_{F_y}^{4_y}$ and $u_{F_y}^{1_y}$ are displacements at points 4 and 1 due to the load F_y, respectively, as shown in Fig. 2.6b), can be called revolute hinges; however, in real-world applications, an ideal revolute hinge can make the center of rotation fixed at point O.

To fulfill the rotation requirement, η should be as small as possible, which means the line determined by points 4 and 1 should cross the geometric center of the flexure hinge, that is, point O (the motion of rigid area **R** is assumed to be rigid motion). Therefore, for a small rotation angle, the following condition should be satisfied:

$$\frac{u_{F_y}^{4_y}}{u_{F_y}^{1_y}} = \frac{\overline{4O}}{\overline{1O}} = \frac{l_d + 2l_s}{l_d}. \tag{2.6}$$

Let

$$\eta^R = \frac{1}{2}\left(\frac{u_{F_y}^{4_y}}{u_{F_y}^{1_y}} - \frac{l_d + 2l_s}{l_d}\right)^2. \tag{2.7}$$

For the topology optimization of revolute hinges, η^R needs to be as small as possible. In conclusion, the optimization model Q_R for revolute hinge topology optimization can be defined as follows:

$$\begin{aligned} Q_R : \min : &\ -C_y + C_x \\ \text{s.t.} : &\ Vol \le V^* \\ &\ \eta^R \le \eta^R_{obj}, \end{aligned} \tag{2.8}$$

where $\eta^R_{obj} \ge 0$ is a given small constant.

2.2.4 Sensitivity Analysis

Here, we use Q_T as an example. By incorporating the SIMP method, the optimization problem Q_T can be rewritten by using the FEM form as follows:

$$\begin{aligned} \min: \quad & J(\boldsymbol{\rho}) = -C_y(\boldsymbol{\rho}) + C_x(\boldsymbol{\rho}) \\ \text{s.t.}: \quad & \sum\nolimits_{e=1}^{N} \rho_e v_e \leq V^* \\ & \eta^T(\boldsymbol{\rho}) \leq \eta^T_{obj} \\ & \boldsymbol{K}\boldsymbol{u}_x = \boldsymbol{F}_x \\ & \boldsymbol{K}\boldsymbol{u}_y = \boldsymbol{F}_y \\ & 0 < \rho_{min} \leq \rho_e \leq 1, e = 1, 2, \ldots, N, \end{aligned} \tag{2.9}$$

where N is the total number of finite elements and F_x and F_y are the load vectors. $\rho_{min} = 0.001$ is introduced to avoid singularity of the stiffness matrix. $\boldsymbol{K}$ is the stiffness matrix in the global sense, which can be written as

$$\boldsymbol{K} = \sum_{e=1}^{N} {\rho_e}^p \boldsymbol{K}_e + \boldsymbol{K}_s, \tag{2.10}$$

where $\boldsymbol{K}_e$ is the (global level) element stiffness matrix and $\boldsymbol{K}_s$ is the stiffness matrix of the output spring k_s at the global level.

The optimization problem can be efficiently solved by the optimality criteria when there is only one constraint [4]. To solve the topology optimization problem with multiple constraints, the method of moving asymptotes (MMA) [35], which belongs to the class of sequential convex programming [2], may be the most computationally efficient approach. Since the optimization problem stated as in Q_T has two constraints in addition to a simple constraint giving upper and lower limits on the density variable, MMA is employed to solve the proposed optimization problem.

The sensitivity of the objective function can be written as

$$\frac{\partial J}{\partial \rho_e} = p\rho_e^{p-1}\boldsymbol{u}_y^{\mathrm{T}}\boldsymbol{K}_e\boldsymbol{u}_y - p\rho_e^{p-1}\boldsymbol{u}_x^{\mathrm{T}}\boldsymbol{K}_e\boldsymbol{u}_x. \tag{2.11}$$

Since the material usage constraint involves only ρ, the sensitivity can be easily obtained and stated as follows:

$$\frac{\partial Vol}{\partial \rho_e} = v_e, \quad e = 1, 2, \ldots N. \tag{2.12}$$

The sensitivity of η^T can be written as

$$\frac{\partial \eta^{\mathrm{T}}}{\partial \rho_e} = (u_{F_y}^{4_y} - u_{F_y}^{1_y})(\frac{\partial u_{F_y}^{4_y}}{\partial \rho_e} - \frac{\partial u_{F_y}^{1_y}}{\partial \rho_e}), \tag{2.13}$$

where

$$\frac{\partial u_{F_y}^{4_y}}{\partial \rho_e} = -p\rho_e^{p-1}\boldsymbol{u}_4^{\mathrm{T}}\boldsymbol{K}_e\boldsymbol{u}_y \tag{2.14}$$

$$\frac{\partial u_{F_y}^{1_y}}{\partial \rho_e} = -p\rho_e^{p-1}\boldsymbol{u}_1^{\mathrm{T}}\boldsymbol{K}_e\boldsymbol{u}_y, \tag{2.15}$$

where u_4 and u_1 are displacement vectors obtained by the load vectors $\boldsymbol{H}_4$ and $\boldsymbol{H}_1$, respectively. $\boldsymbol{H}_4$ and $\boldsymbol{H}_1$ are unit load vectors consisting of zero, except for positions 4 and 1 (see Fig. 2.5), with an entry of $H_4 = 1$ and $H_1 = 1$, respectively. The above method can be directly extended for Q_R.

2.2.5 *Examples*

For the following numerical cases, the artificial material properties are described for a solid material with a Young's modulus $E = 1$ and Poisson's ratio is $\upsilon = 0.3$. The power law mixing rule with $p = 3$ is used in the computations.

2.2.5.1 Translational Hinge Design

In all following cases, four-node isoparametric full integration elements with bilinear shape functions are used, and the applied forces are assumed to be unity. Note that, to clearly plot the optimal topologies, the whole rigid area **R** is not plotted even though different l_s/l_d might be used.

First, the ratio of l_s/l_d is set to 3. The mesh size of **D** is set to100 × 100, and the mesh size of **R** is 100 × 300. The maximum material usage V^* is set to 0.2.

Figure 2.7 shows the obtained optimal configuration. Note that a typical four-bar linkage is obtained. The convergence histories of C_x and C_y are shown in Fig. 2.8. The convergence history of η^T is shown in Fig. 2.9. In this case, η_{obj}^T is set to 0.1, and the resulting η^T is 0.0906. Note that 30 iterations are used in the optimization process. η^T is very large in the second iteration since no reasonable topology is obtained at that point.

Second, we examine the case when l_s/l_d is small, for example. 0.1. The corresponding final design is shown in Fig. 2.10. Note that the optimal flexure hinge is not a simple four-bar linkage. However, if the area in the red circle is treated as a whole, the translational hinge becomes a typical four-bar linkage. The reason for generating the topology is that l_s/l_d is small, which makes the stiffness of the rigid domain **R** small. Therefore, part of the material is added to the rigid domain **R** to

Fig. 2.7 The final design of the translational hinge with $l_s/l_d = 3$

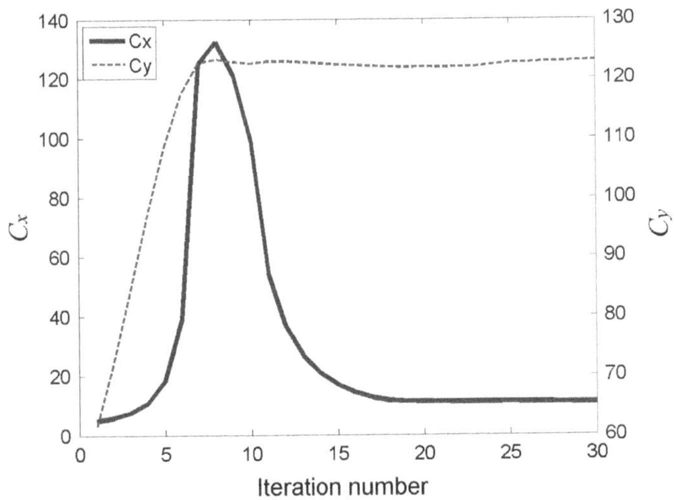

Fig. 2.8 The convergence histories of C_x and C_y with $l_s/l_d = 3$

Fig. 2.9 The convergence history of η^T with $l_s/l_d = 3$

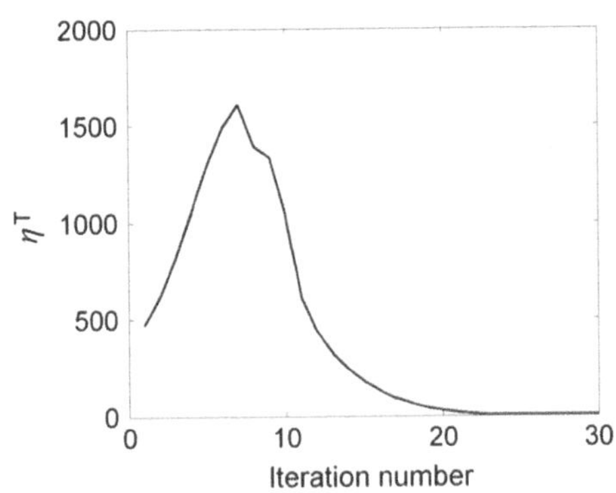

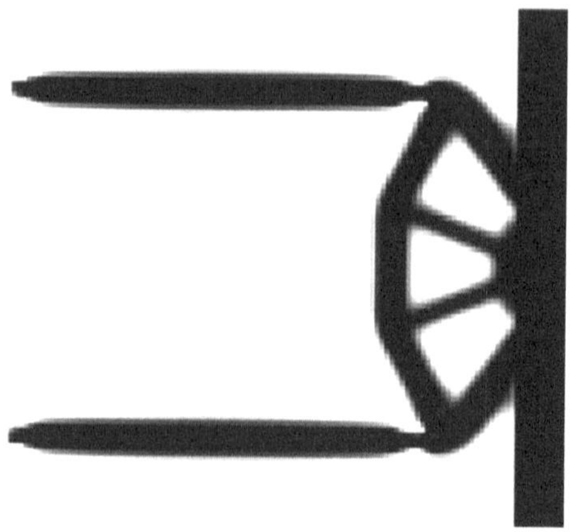

Fig. 2.10 The final design of the translational hinge with $l_s/l_d = 0.1$

make it stiffer. Therefore, when the hinge is loaded, **R** can be moved without torsional deformation. In the case where l_s is small, the problem is actually considered as a compliant mechanism design problem. To categorize the entire design domain as one hinge, l_s generally must be large.

2.2.5.2 Revolute Hinge Design

The relationship between the optimal configuration and the spring k_s is examined in this section. The other conditions are fixed as follows: the mesh sizes of **D** and **R** are set to 100×100 and 100×300, respectively.

The maximum material usage V^* is set to 0.3, and η^R_{obj} is set to 0.001. Four cases are studied in this section [19]: $k_s = 0.01$, $k_s = 0.02$, $k_s = 0.03$ and $k_s = 0.06$.

The corresponding final designs of the four studied cases are shown in Fig. 2.11. Note that the same topology is obtained, although the shape of the final design is slightly different. The shapes of these topologies are different from the shapes of the conventional hinges regardless of whether k_s is large or small. The profile of the obtained flexure hinge is formed by three different segments. The topologies shown in Fig. 2.11 confirm that the topology of the hinges shown in Fig. 2.3 is not the only choice and the best one.

The convergence histories of C_x and C_y are shown in Fig. 2.12. C_x decreases and C_y increases with increasing number of iterations. The convergence history of η^R is shown in Fig. 2.13. One hundred iterations are used in the optimization process. In the case of $k_s = 0.06$, η^R finally approaches 0.001.

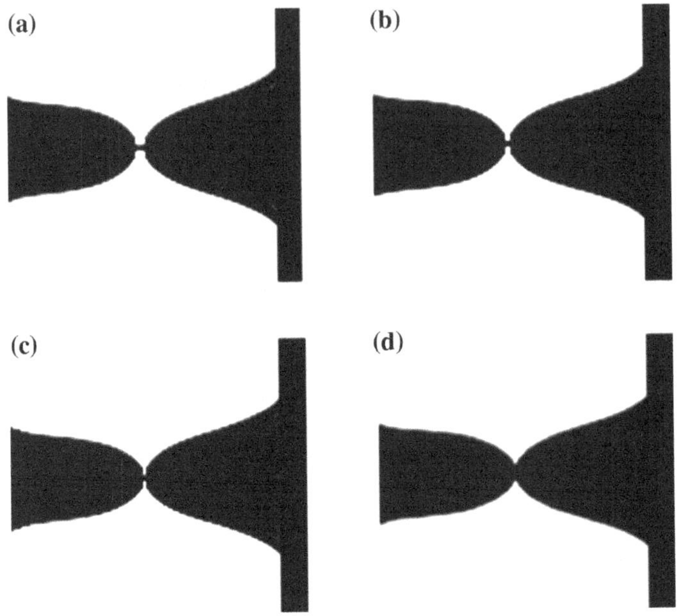

Fig. 2.11 The final designs of revolute hinges with different k_s and the other conditions fixed: **a** $k_s = 0.01$, **b** $k_s = 0.02$, **c** $k_s = 0.03$, and **d** $k_s = 0.06$

Fig. 2.12 The convergence histories of C_x and C_y of the revolute hinge with $k_s = 0.06$

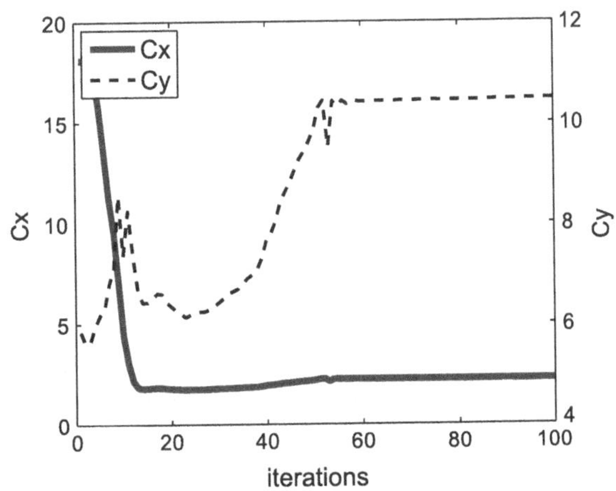

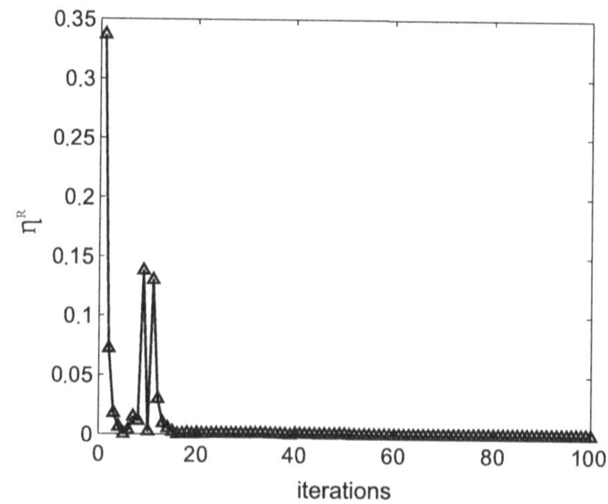

Fig. 2.13 The convergence history of η^R with $k_s = 0.06$

2.3 Stress-Constrained Flexure Hinges

Stress concentration is one of the disadvantages of flexure hinges. It limits the range of motion and reduces the fatigue life of mechanisms. Usually, we expect a flexure hinge to have a large compliance and high rotational precision, but the range of motion and fatigue strength are determined by the maximum stress. Therefore, it is necessary to consider stress levels in the topology optimization of flexure hinges [20].

Flexure hinges are designed to be used in 2D-compliant mechanisms in this chapter. The boundary conditions for hinges are fixed-free: one end is fixed and the other end is free. Since all deformations of flexure hinges are small, the flexure hinge can be modeled and analyzed as a small-displacement fixed-free Euler-Bernoulli beam subjected to bending and axial effects produced by force, moment and axial load. A schematic representation of a hinge with boundary conditions and loads is shown in Fig. 2.14. Points 1, 2, and 3 are the free end, rotation center and fixed end, respectively. Then, the relationship of the displacement load at the free end can be expressed as

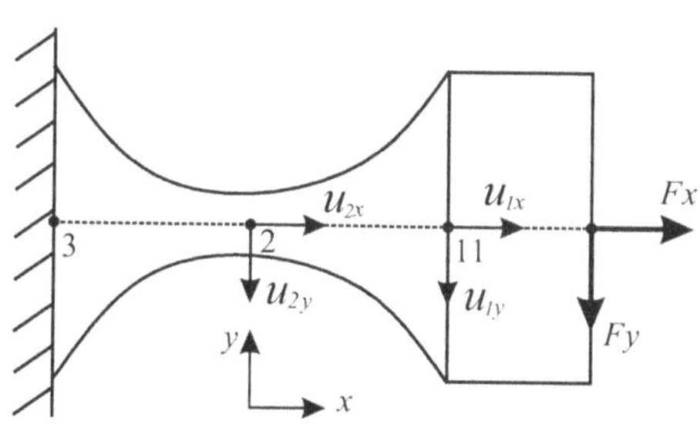

Fig. 2.14 Boundary conditions and loads for determining compliance factors with axial force F_x and transverse F_y

$$X = \mathbf{C}F, \tag{2.16}$$

where $\boldsymbol{X}$ is the displacement vector, $\mathbf{C}$ is the compliance matrix, and $\boldsymbol{F}$ is the load vector. $\boldsymbol{X}$ contains the rotational displacement θ, and translation displacements u_x and u_y are in the x- and y-directions, respectively. The load vector $\boldsymbol{F}$ contains axial force F_x, transverse force F_y, and bending moment M_z. The detailed expression is shown in Eq. (2.17).

$$\begin{bmatrix} \theta \\ u_y \\ u_x \end{bmatrix} = \begin{bmatrix} C_{11} & C_{12} & 0 \\ C_{21} & C_{22} & 0 \\ 0 & 0 & C_{33} \end{bmatrix} \begin{bmatrix} M_z \\ F_y \\ F_x \end{bmatrix} \tag{2.17}$$

where C_{ij} are the compliance factors, and $C_{12} = C_{21}$ according to the reciprocity principle.

Figure 2.5 shows the flexure hinge design domain. Ω_d is set to be a square and $l_s = 3l_d$. An ideal flexure hinge should be as compliant as possible in the desired direction while being as stiff as possible in all other directions. Therefore, compliance factors C_{11}, C_{12}, and C_{22} need to be maximized while C_{33} needs to be minimized. To simplify the optimization modeling, compliance factors C_{22} due to force F_y and C_{33} due to force F_x are selected as the objectives; therefore, we need to maximize C_{22} and minimize C_{33}.

Based on Eq. (2.17), compliance factors C_{22} and C_{33} can be written as

$$C_{22} = \frac{u_{1y}^{F_y}}{F_y}, \quad C_{33} = \frac{u_{1x}^{F_x}}{F_x}, \tag{2.18}$$

where $u_{1x}^{F_x}$ is the displacement of point 1 along the x-direction due to F_x, and $u_{1y}^{F_y}$ is the displacement of point 1 along the y-direction due to F_y.

The objective function contains two subgoals, C_{22} and C_{33}, which are addressed using a dimensionless method. The multi-objective optimization is converted into a single-objective optimization problem by using the weighted sum method. The objective function can be expressed as follows:

$$\min: \quad -w\frac{C_{22}}{C_{22}^0} + (1-w)\frac{C_{33}}{C_{33}^0}, \tag{2.19}$$

where w $(0 \le w \le 1)$ is the weighting factor and C_{22}^0 and C_{33}^0 are the initial values in the topology optimization.

To improve the rotational precision, the following constraint function [46] is used to limit the distance d (Fig. 2.5)

$$\xi = \left(\frac{u_{4y}^{F_y}}{u_{1y}^{F_y}} - \frac{l_d + 2l_s}{l_d} \right)^2 \le \xi^*, \tag{2.20}$$

where $u_{4y}^{F_y}$ is the displacement of input port 4 along the y-direction due to F_y, and ξ^* is a given small positive constant.

In addition, flexure hinges must satisfy the material yield strength requirement. The maximum stress cannot exceed the stress limit. Therefore, the stress constraint is added to the optimization problem

$$\sigma_{max} = \max(\boldsymbol{\sigma}) \leq \sigma^*, \tag{2.21}$$

where σ_{max} is the maximum stress, and σ^* is a chosen threshold for the maximum elemental stress. The vector $\boldsymbol{\sigma}$ contains the element stress measures, which will be defined in Sect. 1.3.1.

The general formulation for the topology optimization of flexure hinges is given as

$$\begin{aligned}
\min: \quad & -w\frac{C_{22}}{C_{22}^0} + (1-w)\frac{C_{33}}{C_{33}^0} \\
\text{s.t}: \quad & \boldsymbol{v}^T\boldsymbol{\rho} \leq V^* \\
& \xi \leq \xi^* \\
& \sigma_{max} \leq \sigma^* \\
& 0 \leq \rho_i \leq 1, i = 1, 2, \ldots, N \\
& \textit{Equilibrium equations}
\end{aligned} \tag{2.22}$$

where $\boldsymbol{v}$ is an N-vector containing the element volumes, and $\boldsymbol{\rho}$ is the vector of design variables.

2.3.1 Stress Criterion

Generally, the designs are required to pass material failure criteria, which are based on the stress tensor in most engineering situations. Therefore, the applicability of the topology optimization method can be greatly improved when a stress constraint is added to the optimization model. However, stress-based topology optimization remains challenging.

Stress-constrained topology optimization was first encountered for the optimization problem of truss structures, and it was noticed by Kirsch [13]. The optimization problem with truss was overcome by the *ε–relaxation* method [7] and smooth envelope functions [28]. Later, Bruggi [5] introduced these approaches for the stress-constrained design of continuum structures. In recent years, Le et al. [15] presented a normalized global/regional stress measure based on the well-know *P-norm* of the von Mises effective stress. Holmberg et al. [10] proposed a clustering technique. Based on a modified *P-norm*, the stress evaluation points were clustered into several groups to decrease the number of stress constraints and the computational cost. De Leon et al. [9] used the relaxation scheme introduced by Le et al. [15] to design compliant mechanisms, and the one-node hinges were alleviated by the stress constraint. In this paper, a global stress measure with an adaptive normalization scheme [15] is applied to the von Mises stress on each design element.

The element stress vector $\boldsymbol{\sigma}_i$ of a fully solid material at the stress evaluation point is written as

$$\boldsymbol{\sigma}_i = [\sigma_{11}, \sigma_{22}, \sigma_{12}]^T. \tag{2.23}$$

It is calculated in finite element analysis as

$$\boldsymbol{\sigma}_i = \mathbf{DB}\boldsymbol{u}_i, \tag{2.24}$$

where $\mathbf{D}$ is the constitutive matrix for the base material, $\mathbf{B}$ is the strain-displacement matrix, and $\boldsymbol{u}_i$ is the displacement vector of the ith element.

The von Mises equivalent stress measure is often applied for statically loaded structures; therefore, it is used as the stress measure in the topology optimization of flexure hinges. The expression for the von Mises equivalent stress in the case of plane stress is as follows:

$$\begin{aligned} \sigma_i^{vM} &= \sqrt{\sigma_{11}{}^2 + \sigma_{22}{}^2 + 3\sigma_{12}{}^2 - \sigma_{11}\sigma_{22}} \\ &= \left(\boldsymbol{\sigma}_i^T \mathbf{V} \boldsymbol{\sigma}_i\right)^{1/2}, \end{aligned} \tag{2.25}$$

where $\mathbf{V}$ is the auxiliary matrix:

$$\mathbf{V} = \begin{bmatrix} 1 & -1/2 & 0 \\ -1/2 & 1 & 0 \\ 0 & 0 & 3 \end{bmatrix} \tag{2.26}$$

To allow the element stress measure to go to zero when ρ_i is close to 0 and to recover the original stress when ρ_i is close to 1, the solid material stresses are also penalized for intermediate design variable values. The element stress measure for a given element becomes

$$\sigma_i = \rho_i{}^q \sigma_i^{vM}, \tag{2.27}$$

where σ_i is the relaxed stress measure of the ith element and q is a relaxation parameter. A value of $q = 0.5$ is used to generate the topology results.

In finite element-based topology optimization, the design domain is discretized into N finite elements. Each element has a stress constraint. This implies that the number of stress constraints N is very large, and the sensitivity computation is prohibitively costly. The most common way to reduce the computational burden is to use one global stress constraint instead of the N local stress measures, thereby improving computational efficiency. Therefore, the global *P-norm* stress measure is adopted to control the maximum stress in this work. The *P-norm* stress measure can be written as:

$$\tilde{\sigma}_{pn} = \left(\sum_{i=1}^{N} v_i \sigma_i{}^P \right)^{\frac{1}{P}}, \tag{2.28}$$

where v_i is the element volume, and P is the stress norm parameter. When $P \to \infty$, *P-norm* approaches the maximum stress, but a higher value of P is not necessarily better. Higher values of P are accompanied by a greater degree of non-linearity and discontinuity. Based on the authors' experience, $P = 12$ is a valid choice.

The normalized global stress measure proposed by Le et al. [15] is used to better approximate the maximum stress

$$\sigma_{max} \approx \sigma_{pn} = c\tilde{\sigma}_{pn} \tag{2.29}$$

where c is the normalization parameter calculated at each optimization iteration $n \geq 1$ as follows:

$$c^n = \alpha^n \frac{\sigma_{max}^{n-1}}{\tilde{\sigma}_{pn}^{n-1}} + (1 - \alpha^n)c^{n-1}, \tag{2.30}$$

where σ_{max}^{n-1} is the actual maximum element stress and $\tilde{\sigma}_{pn}^{n-1}$ is the *P-norm* stress from the previous optimization iteration $n - 1$. $\alpha^n \in (0, 1]$ is a parameter that controls the update of c between iterations. In this paper, $c^0 = 1$ and $\alpha^n = 0.5$ for all n.

Flexure hinges are mainly subjected to bending loads, and the stress produced by axial load F_x is relatively small when compared with the stress caused by the force F_y. Therefore, the maximum stress caused by the axial force is neglected in this paper, and the stress constraint of the flexure hinge can be formulated as:

$$\sigma_{max}^{F_y} \approx \sigma_{pn}^{F_y} \leq \sigma^*, \tag{2.31}$$

where $\sigma_{max}^{F_y}$ and $\sigma_{pn}^{F_y}$ are the maximum stress and normalized *P-norm* stress measure due to force F_y, respectively.

2.3.2 Sensitivity Analysis

The adjoint sensitivity method is applied to calculate the gradients. The finite element discrete form for the stress-constrained optimization problem of flexure hinges can be formulated as:

$$
\begin{aligned}
\min: \quad & f(\boldsymbol{\rho}) = -w\frac{C_{22}}{C_{22}^0} + (1-w)\frac{C_{33}}{C_{33}^0} \\
\text{s.t}: \quad & \mathbf{K}\boldsymbol{u}_x = \boldsymbol{F}_x \\
& \mathbf{K}\boldsymbol{u}_y = \boldsymbol{F}_y \\
& f_v(\boldsymbol{\rho}) = \boldsymbol{v}^T\boldsymbol{\rho} \le V^* \\
& f_\xi(\boldsymbol{\rho}) = \xi \le \xi^* \\
& f_s(\boldsymbol{\rho}) = \sigma_{max}^{F_y} \le \sigma^* \\
& 0 \le \rho_i \le 1, i = 1, 2, \ldots, N
\end{aligned}
\tag{2.32}
$$

where $\mathbf{K}$ is the global stiffness matrix, and vectors $\boldsymbol{u}_x$ and $\boldsymbol{u}_y$ are the displacement fields due to load vectors $\boldsymbol{F}_x$ and $\boldsymbol{F}_y$, respectively. N is the total number of finite elements.

Assuming design-independent loads, the sensitivity of the objective function can be written as:

$$
\frac{\partial f}{\partial \rho_i} = -\frac{w}{C_{22}^0 F_y}\frac{\partial u_{1y}^{F_y}}{\partial \rho_i} + \frac{1-w}{C_{33}^0 F_x}\frac{\partial u_{1x}^{F_x}}{\partial \rho_i},
\tag{2.33}
$$

where

$$
\begin{aligned}
\frac{\partial u_{1y}^{F_y}}{\partial \rho_i} &= -p{\rho_i}^{p-1}(\boldsymbol{u}_1^y)^T\mathbf{K}_0\boldsymbol{u}_y \\
\frac{\partial u_{1x}^{F_x}}{\partial \rho_i} &= -p{\rho_i}^{p-1}(\boldsymbol{u}_1^x)^T\mathbf{K}_0\boldsymbol{u}_x,
\end{aligned}
\tag{2.34}
$$

where $\boldsymbol{u}_1^y$ and $\boldsymbol{u}_1^x$ are displacement vectors obtained by the adjoint load vectors $\boldsymbol{L}_1^y$ and $\boldsymbol{L}_1^x$, respectively. $\boldsymbol{L}_1^x$ and $\boldsymbol{L}_1^y$ are load vectors with unit loads in the x- and y-directions at point 1, respectively, and $\mathbf{K}_0$ is the element stiffness matrix.

The sensitivity of the volume constraint can be easily derived:

$$
\frac{\partial f_v}{\partial \rho_i} = v_i, \quad i = 1, 2, \ldots, N.
\tag{2.35}
$$

The sensitivity of f_ξ can be written as follows:

$$
\frac{\partial f_\xi}{\partial \rho_i} = 2\left(\frac{u_{4y}^{F_y}}{u_{1y}^{F_y}} - \frac{l_d + 2l_s}{l_d}\right)\frac{\frac{\partial u_{4y}^{F_y}}{\partial \rho_i}u_{1y}^{F_y} - \frac{\partial u_{1y}^{F_y}}{\partial \rho_i}u_{4y}^{F_y}}{\left(u_{1y}^{F_y}\right)^2},
\tag{2.36}
$$

where

$$
\frac{\partial u_{4y}^{F_y}}{\partial \rho_i} = -p{\rho_i}^{p-1}(\boldsymbol{u}_4^y)^T\mathbf{K}_0\boldsymbol{u}_y,
\tag{2.37}
$$

where $\boldsymbol{u}_4^y$ is the displacement vector obtained by the load vector $\boldsymbol{L}_4^y$, which is a unit load vector consisting of zero except for the degree of freedom in the y-direction at input port 4, with an entry of $\boldsymbol{L}_4^y = 1$.

The stress constraint is the *P-norm* stresses in Eq. (2.31), and the gradients follow from the chain rule:

$$\begin{aligned}\frac{\partial f_s}{\partial \rho_i} &= \frac{\partial \sigma_{pn}}{\partial \rho_i} \\ &= \frac{\partial \sigma_{pn}}{\partial \sigma_i}\left[q{\rho_i}^{q-1}\sigma_i^{vM} + {\rho_i}^q\left(\frac{\partial \sigma_i^{vM}}{\partial \boldsymbol{\sigma}_i}\right)^T \frac{\partial \boldsymbol{\sigma}_i}{\partial \rho_i}\right].\end{aligned} \tag{2.38}$$

The derivatives in Eq. (2.38) are calculated in the following subsections.

The $\partial \sigma_{pn}/\partial \sigma_i$ term in Eq. (2.38) is derived as follows:

$$\begin{aligned}\frac{\partial \sigma_{pn}}{\partial \sigma_i} &= c\frac{1}{P}\left(\sum_{i=1}^{N} v_i {\sigma_i}^P\right)^{\frac{1}{P}-1} v_i P {\sigma_i}^{P-1} \\ &= c\left(\sum_{i=1}^{N} v_i {\sigma_i}^P\right)^{\frac{1}{P}-1} v_i {\sigma_i}^{P-1}\end{aligned} \tag{2.39}$$

The derivatives of the von Mises stress with respect to its stress components are

$$\begin{aligned}\frac{\partial \sigma_i^{vM}}{\partial \sigma_{11}} &= \frac{1}{2\sigma_i^{vM}}(2\sigma_{11} - \sigma_{22}) \\ \frac{\partial \sigma_i^{vM}}{\partial \sigma_{22}} &= \frac{1}{2\sigma_i^{vM}}(2\sigma_{22} - \sigma_{11}) \\ \frac{\partial \sigma_i^{vM}}{\partial \sigma_{12}} &= \frac{3}{\sigma_i^{vM}}\sigma_{12}\end{aligned} \tag{2.40}$$

The derivative of the stress vector with respect to design variable ρ_i is stated as:

$$\frac{\partial \boldsymbol{\sigma}_i}{\partial \rho_i} = \mathbf{DB}\frac{\partial \boldsymbol{u}_i}{\partial \rho_i} \tag{2.41}$$

The adjoint method is used to solve the term $\partial \boldsymbol{u}_i/\partial \rho_i$ in Eq. (2.41). It is calculated from the global equilibrium equation:

$$\frac{\partial \mathbf{K}}{\partial \rho_i}\boldsymbol{u}_y + \mathbf{K}\frac{\partial \boldsymbol{u}_y}{\partial \rho_i} = \mathbf{0} \tag{2.42}$$

So the term $\partial \boldsymbol{u}_i/\partial \rho_i$ can be obtained for each element:

$$\frac{\partial \boldsymbol{u}_i}{\partial \rho_i} = -\mathbf{K}^{-1}\frac{\partial \mathbf{K}}{\partial \rho_i}\boldsymbol{u}_i \tag{2.43}$$

Substituting Eq. (2.43) into Eq. (2.41) and then substituting Eqs. (2.41) and (2.39) into Eq. (2.38), the gradient is obtained as

$$\begin{aligned} \frac{\partial f_s}{\partial \rho_i} &= \frac{\partial \sigma_{pn}}{\partial \rho_i} \\ &= c\left(\sum_{i=1}^{N} v_i {\sigma_i}^P\right)^{\frac{1}{P}-1} v_i {\sigma_i}^{P-1} q {\rho_i}^{q-1} \sigma_i^{vM} - {\boldsymbol{\lambda}_i}^T \frac{\partial \mathbf{K}}{\partial \rho_i}\boldsymbol{u}_i, \end{aligned} \tag{2.44}$$

where an adjoint variable $\boldsymbol{\lambda}_i$ is defined as:

$$\boldsymbol{\lambda}_i{}^T = c\left(\sum_{i=1}^{N} v_i {\sigma_i}^P\right)^{\frac{1}{P}-1} v_i {\sigma_i}^{P-1} {\rho_i}^q \left(\frac{\partial \sigma_i^{vM}}{\partial \boldsymbol{\sigma}_i}\right)^T \mathbf{DBK}^{-1}, \tag{2.45}$$

which means that it can be derived from the adjoint equation:

$$\mathbf{K}\boldsymbol{\lambda}_i = c\left(\sum_{i=1}^{N} v_i {\sigma_i}^P\right)^{\frac{1}{P}-1} v_i {\sigma_i}^{P-1} {\rho_i}^q \mathbf{B}^T\mathbf{D}^T \frac{\partial \sigma_i^{vM}}{\partial \boldsymbol{\sigma}_i}. \tag{2.46}$$

2.3.3 Examples

The design domain is shown in Fig. 2.5, where (Ω_d) is fixed on the left edge. The rigid domain (Ω_s) is a group of elements that are inactive, which means they are forced to have density equal to 1 and are ignored in the calculations of the global stress measure. Two point loads, F_x and F_y, 50 N and 1 N, respectively, are applied at the input port. The point load generates high local stress, which would overshadow other highly stressed regions. Therefore, the finite elements near the point load should be excluded from the design variables: they are forced to have density equal to 1 and are ignored in the optimization problem. The high local stress produced by a point load does not need to be considered since the loads are active in the non-design domain (Ω_s), which is excluded from the optimization problem.

Based on the linear finite element analysis (FEA), the plane stress FEA uses bilinear four-node quadrilateral elements with a thickness of 1 mm. The artificial material used in the model has a Young's modulus of $E = 1$ GPa and a Poison's ratio of $\mu = 0.3$. The lengths of the design and rigid domains are set to $l_d = 10$ mm and $l_s = 30$ mm, respectively. The weighting factor is $w = 0.5$, the volume constraint is $V^* = 0.3$, ξ^* is set to 0.001, and the stress limit is stated for each result.

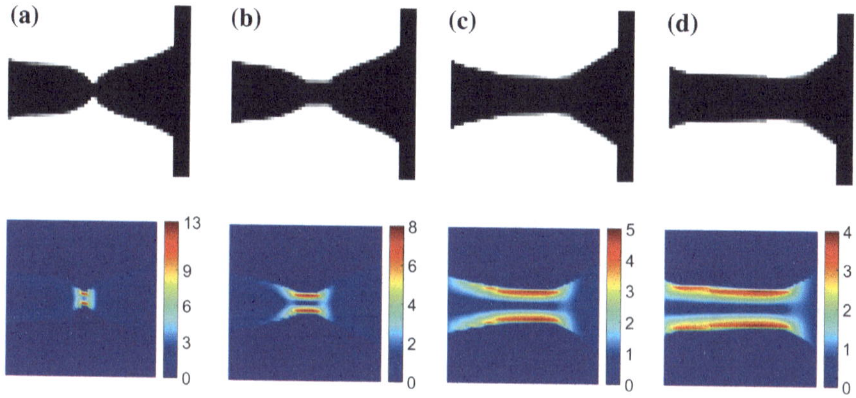

Fig. 2.15 Optimized topologies (top) and stress distributions (bottom) for a 50 × 50 mesh with various stress constraint limits: **a** without stress constraint $\sigma_{max} = 13.2$; **b** $\sigma^* = 8$; **c** $\sigma^* = 5$; **d** $\sigma^* = 4$. All the values of σ_{max} and σ^* are in MPa

2.3.3.1 Effects of Mesh Refinement and Stress Constraint

To investigate the effects of mesh refinement and stress constraint on the topology results of flexure hinges, two examples with meshes of 50 × 50 and 100 × 100 are presented. Several stress constraints are applied in both examples, and the output spring stiffness is $K_{out} = 0.008$ N/mm.

Figure 2.15 shows the topology results and the stress distributions for a 50 × 50 mesh with several levels of stress limit. Figure 2.15a is the final topology result without stress constraint, and the maximum stress is 13.2 MPa. Some sharp corners are present in the middle segment of the flexure hinge, which indicates that stress concentration is easy to produce. The topology results with stress limits are shown in Fig. 2.15b–d. The stress constraints effectively control the local stress and reduce the maximum stress; additionally, the length and thickness of the middle segment increase. For a very low value, $\sigma^* = 4$ (Fig. 2.15d), the topology result exhibits distributed stress rather than local stress. The hinge is similar to a corner-filleted flexure hinge [23], which confirms that the corner-filleted flexure hinge has a low stress level.

The optimized topologies and the stress distributions for a 100 × 100 mesh are shown in Fig. 2.16. Figures 2.16a–d show the topologies without stress constraint and with stress limits of 8, 5, and 4, respectively. As the stress limit decreases, the maximum stress decreases and the topology changes. The most significant change is the removal of the sharp corners and the smoothing of the profile of the middle segment, which means that the stress constraint effectively reduces the stress concentration.

The topologies without stress constraints (Figs. 2.15a and 2.16a) are nearly the same, which means that the topology does not change with mesh refinement when a stress constraint is not added. However, the topology results vary with mesh refinement when a stress constraint is added. As the mesh is refined, the middle segment

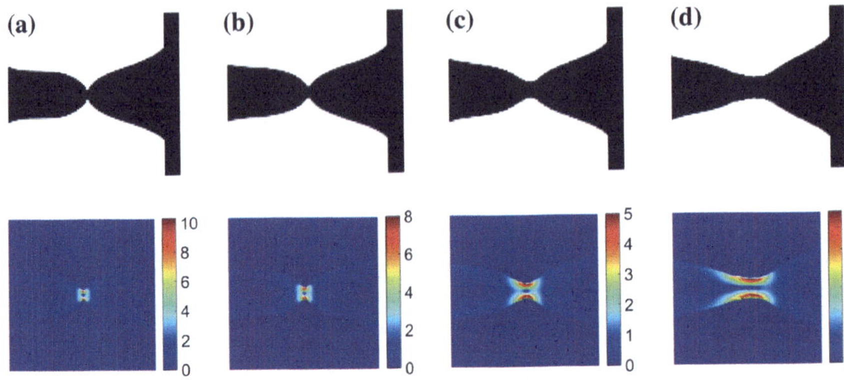

Fig. 2.16 Optimized topologies (top) and stress distributions (bottom) for a 100 × 100 mesh with various stress constraint limits: **a** without stress constraint $\sigma_{max} = 10.3$; **b** $\sigma^* = 8$; **c** $\sigma^* = 5$; **d** $\sigma^* = 4$. All the values of σ_{max} and σ^* are in MPa

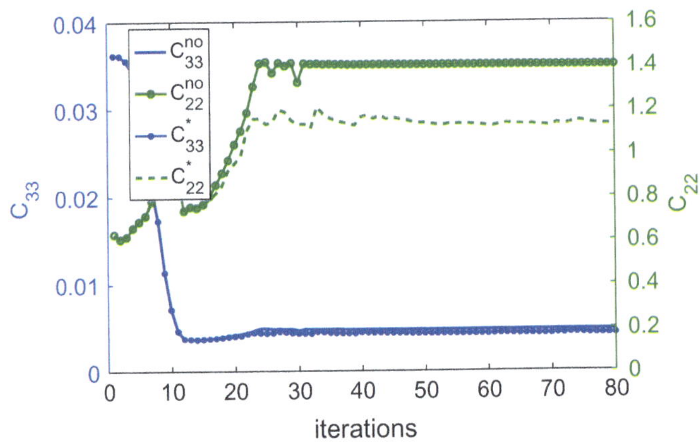

Fig. 2.17 The convergence histories of the objective function without (Fig. 2.16a) and with a stress limit of $\sigma^* = 5$ (Fig. 2.16c)

becomes thinner and smoother rather than the entire hinge becoming thicker and longer, demonstrating that mesh refinement affects the topology in stress-constrained topology optimization of flexure hinges.

Note that the maximum stress in Fig. 2.16a is 10.3 MPa (less than 13.2 MPa). The mesh refinement does not lead to stress reduction, but the result of the FEA is different for different meshing densities.

Figures 2.17 and 2.18 show the convergence histories of the objective function and the *P-norm* stress measure for the 100 × 100 mesh, respectively. The superscript (^{no}) indicates the topology result without stress constraint, and the superscript (*) indicates the topology result with a stress limit of $\sigma^* = 5$. C_{22}^{no} and C_{22}^* gradually increase and C_{33}^{no} and C_{33}^* gradually decrease with increasing number of iterations.

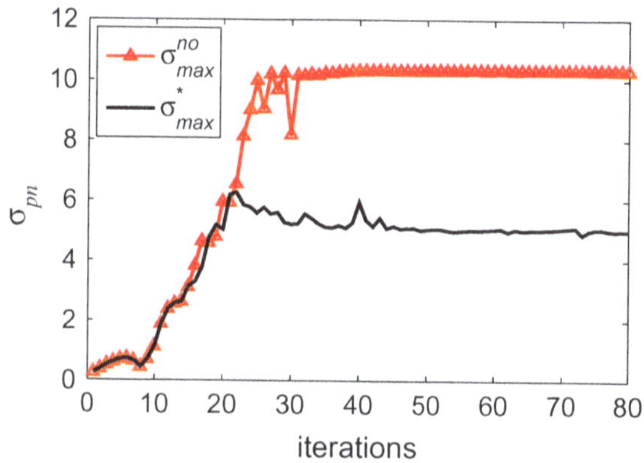

Fig. 2.18 The convergence histories of the *P-norm* stress measure without (Fig. 2.16a) and with a stress limit of $\sigma^* = 5$ (Fig. 2.16c)

The maximum stress σ^{no}_{max} also increases with increasing number of iterations. When the stress constraint is added to the optimization model, the maximum stress σ^{*}_{max} converges to the prescribed stress limit, and the optimized topology is obtained after eighty iterations.

2.3.3.2 Effect of Output Spring Stiffness K_{out}

To further investigate the relationship between the optimized topology and the output stiffness K_{out}, several output stiffnesses are considered in the optimization process.

Figure 2.19 shows the topology results for different output springs, $K_{out} = 0.004$, 0.003, 0.002 and 0.001, both without and with a stress limit of $\sigma^* = 8$. Figure 2.19a, c, e and g show that the middle segment of the flexure hinge becomes very thin with decreasing of output stiffness. The flexure hinge consists of a narrow rectangular region with thicker regions on either side, which is like a small-length flexural pivot. The deformation of the hinge mainly occurs in this section, and the maximum stress is very large. The thickness and length of the middle segment vary substantially when the stress constraint is added (as shown in the bottom of Fig. 2.19). For $K_{out} = 0.004$, the stress constraint effectively reduces the maximum stress and smoothes the middle segment. For $K_{out} = 0.003$ and 0.002, stress constraint makes the middle segment become slender. Although the maximum stress is reduced, the hinge is similar to a constant rectangular cross section flexure hinge. For very low $K_{out} = 0.001$, a thick circular member is formed in the middle section, which makes the deformation occur at either side of the circular member. This may be because the output stiffness is too small, resulting in convergence to a local Pareto-optimal front. Therefore, the output spring stiffness cannot be too small for topology optimization of flexure hinges.

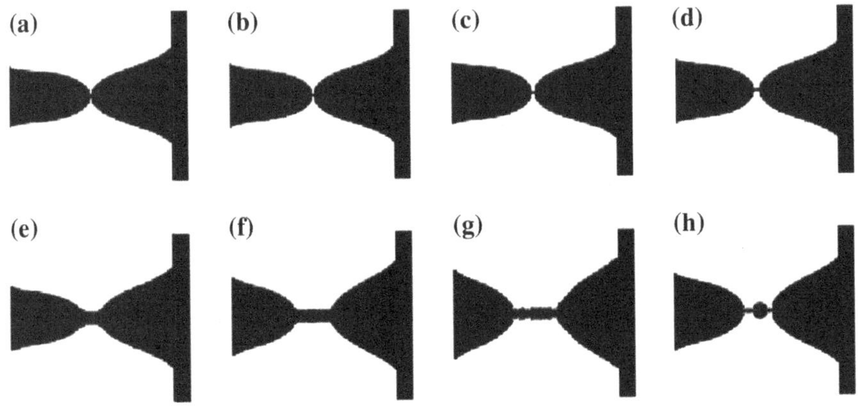

Fig. 2.19 Optimized topologies without (top) and with stress constraints for different output stiffnesses. The units for all examples are N/mm for K_{out} and MPa for σ_{max} and σ^*. **a** $K_{out} = 0.004$, $\sigma_{max} = 12.1$, **b** $K_{out} = 0.004$, $\sigma^* = 8$, **c** $K_{out} = 0.003$, $\sigma_{max} = 17.1$, **d** $K_{out} = 0.003$, $\sigma^* = 8$, **e** $K_{out} = 0.002$, $\sigma_{max} = 19.5$, **f** $K_{out} = 0.002$, $\sigma^* = 8$, **g** $K_{out} = 0.001$, $\sigma_{max} = 20.8$ and **h** $K_{out} = 0.001$, $\sigma^* = 8$

2.3.3.3 Comparison of Flexure Hinges With and Without Stress Constraints

The maximum stress of a flexure hinge is examined using ANSYS. According to the topology results (Fig. 2.16a and c), two flexure hinges are modeled, and the performances of the two hinges are compared. The material of the flexure hinges is an aluminum alloy, which has a Young's modulus $E = 71$ GPa and a Poisson's ratio $\mu = 0.33$. l_d and l_s are set to 10 and 30 mm, respectively, and the two hinges have the same cross-sectional width $b = 10$ mm.

Since the minimum thickness of the two flexure hinges is different, the stress concentration factor is calculated and compared. For an applied force F_y, the maximum stress at the notch region can be calculated as:

$$\sigma_{max} = k_b \sigma_{nom} = k_b \frac{6F_y[l_s + (1/2)l_d]}{bt^2}, \tag{2.47}$$

where σ_{nom} is the nominal stress, t is the minimum thickness and k_b is the stress concentration factor. k_b can be calculated from Eq. (2.48) as

$$k_b = \frac{bt^2 \sigma_{max}}{6F_y[l_s + (1/2)l_d]}. \tag{2.48}$$

The displacements of points 1 and 4 along the y-direction, the maximum stress and the stress concentration factor are shown in Table 2.1. F5(a) and F5(c) denote the hinges obtained from Fig. 2.16a and c, respectively.

Table 2.1 Comparison of the displacement (mm), maximum stress (MPa) and stress concentration factor for two flexure hinges

	$u_4^y(\times 10^{-2})$	$u_1^y(\times 10^{-3})$	σ_{nom}	σ_{max}	k_b
F5(a)	5.972	8.776	60.69	67.96	1.119
F5(b)	3.432	5.087	24.73	25.77	1.042
F5(c)	5.972	8.920	42.34	44.25	1.045

In the first comparison, the same input load $F_y = 1$ N is applied at input port 4 (as shown in Fig. 2.5). The first and second rows of Table 2.1 indicate that for the same maximum material usage, the maximum stress and stress concentration factor of the flexure hinge with stress constraint are less than those of the flexure hinge without stress constraint.

In the second comparison, the same input displacement is applied at the input port for the two hinges. The first and third rows of Table 2.1 show that the hinge with stress constraint also has a lower stress level than the hinge without stress constraint, which confirms that the stress constraint effectively reduces the stress concentration problem of flexure hinges.

2.4 Redesign of Notch Flexure Hinges

The performance of a flexure hinge is determined by its profile. The aforementioned studies of flexure hinges have used numerous notch profiles to improve the performance of such hinges. The performance of flexure hinges is primarily evaluated with respect to three criteria: compliance, precision of rotation, and stress levels. An ideal flexure hinge should have large compliance in the rotational direction, large stiffness in the direction of parasitic motion, high precision, and low stress levels. However, simultaneously satisfying all these requirements is difficult due to the elastic deformation of flexure hinges. Xu and King [42] investigated and compared the performances of right-circular, corner-filleted and elliptical flexure hinges in terms of compliance, precision and stress concentration. The corner-filleted flexure hinges had the largest compliance but lower rotational precision, the elliptical flexure hinges had low stress, and the right-circular flexure hinges had the highest rotational precision but small compliance. The results indicated that a flexure hinge generally has only one or two optimal performances. Right-circular and V-shaped flexure hinges [36], which have a thin section in the middle section and bulk on either side, possess high rotational precision but small compliance due to the bulk. One possible approach to increase the compliance while simultaneously retaining high precision is to remove some material from the flexure hinges. In other words, we can change the topology of the notch flexure hinges, thereby changing their performance.

However, it is very difficult to optimally remove material from flexure hinges, and the removal process depends on the designers' experience. This problem can be solved by using topology optimization, which can automatically generate the optimal geometries.

2.4.1 Design Domain Setting

This section is devoted to developing an optimization model for removing material from flexure hinges [21]. As mentioned above, the shape of the profile itself determines the characteristics of the flexure hinge, such as the compliance, rotational precision and allowed range of rotation. The right-circular flexure hinge is selected in this study since it has higher precision of rotation than that of other conventional flexure hinges, such as the leaf-type and corner-filleted hinges. Although right-circular flexure hinges are precise with respect to rotation, i.e., the center of rotation does not deviate from the initial position as much as other hinges, the compliance and range of motion of right-circular flexure hinges are small, especially when the t/R ratio (as shown in Fig. 2.20a) is relatively large. Therefore, we attempt to remove some of the material from the right-circular flexure hinge to improve its performance. A schematic representation of a right-circular flexure hinge without and with material removal is shown in Fig. 2.20. However, as mentioned previously, it is difficult to determine the position, shape, size and number of holes. The topology optimization approach can effectively overcome these issues. The advantages of topology optimization lie in the fact that the choice of the appropriate topology of a structure or a mechanism in the conceptual phase is generally the most decisive factor in the efficiency of a novel product.

Figure 2.21 shows the design domain of a flexure hinge, where Ω_d is the design domain, which has a prescribed profile of a right-circular flexure hinge; Ω_s is a prescribed solid domain (element densities are set to 1); and Γ_D is the Dirichlet boundary condition with prescribed displacements. The design domain is fixed at

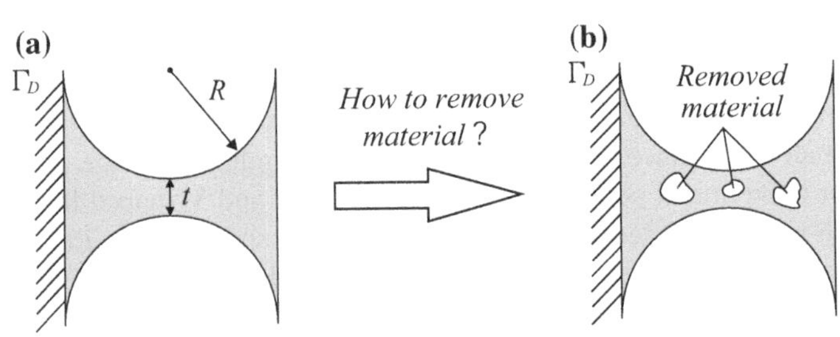

Fig. 2.20 A sketch of material removal from a right-circular flexure hinge: **a** without and **b** with material removal. In the figure, Γ_D is the boundary condition, t is the minimum thickness, and R is the radius

the left end, and point 2 is the input port, which is subjected to loads F_x and F_y, where h ($h = t + 2R$) is the height of the design domain and l_s is the length of the solid domain. In addition, artificial springs k_{in} and k_{out} are added to points 2 and 1, respectively. The spring in the y-direction makes the optimization problem a well-posed design problem, thereby avoiding numerical instability, and the spring in the x-direction makes the flexure hinge exhibit a degree of elasticity.

2.4.2 Optimization Models

Generally, the compliance of point 1 (see Fig. 2.21) represents the compliance of the flexure hinge. The flexure hinge should be as compliant as possible in the intended degrees of freedom and as stiff as possible in the directions of parasitic motion. Under the given constraint conditions, such as the prescribed material volume fraction, the compliances in the y- and x-directions need to be maximized and minimized, respectively. In other words, we need to maximize the displacement of point 1 due to F_y along the y-direction and minimize the displacement of point 1 due to F_x along the x-direction. The objective function adopts a dimensionless formulation, and the general formulation for the optimization model Θ_1 is given as:

$$
\begin{aligned}
\Theta_1: \quad \min: \quad & f(\boldsymbol{\rho}) = \frac{u_{1x}(\boldsymbol{\rho})/u_{1x}^0}{u_{1y}(\boldsymbol{\rho})/u_{1y}^0} \\
\text{s.t}: \quad & \boldsymbol{K}\boldsymbol{u}_x = \boldsymbol{F}_x \\
& \boldsymbol{K}\boldsymbol{u}_y = \boldsymbol{F}_y \\
& f_v(\boldsymbol{\rho}) = \frac{\boldsymbol{v}^T\boldsymbol{\rho}}{Vol_0} \le V^* \\
& 0 < \rho_{min} \le \rho_i \le 1, i = 1, 2 \cdots, N
\end{aligned}
\tag{2.49}
$$

where $u_{1x}(\boldsymbol{\rho})$ is the displacement of point 1 along the x-direction due to F_x; $u_{1y}(\boldsymbol{\rho})$ is the displacement of point 1 along the y-direction due to F_y; u_{1x}^0 and u_{1y}^0 are the initial values in the topology optimization; $\boldsymbol{F}_x$ and $\boldsymbol{F}_y$ are vectors, which take zero in all positions except for the position corresponding to the input degree of freedom; and vectors $\boldsymbol{u}_x$ and $\boldsymbol{u}_y$ are the displacement fields due to load vectors $\boldsymbol{F}_x$ and $\boldsymbol{F}_y$,

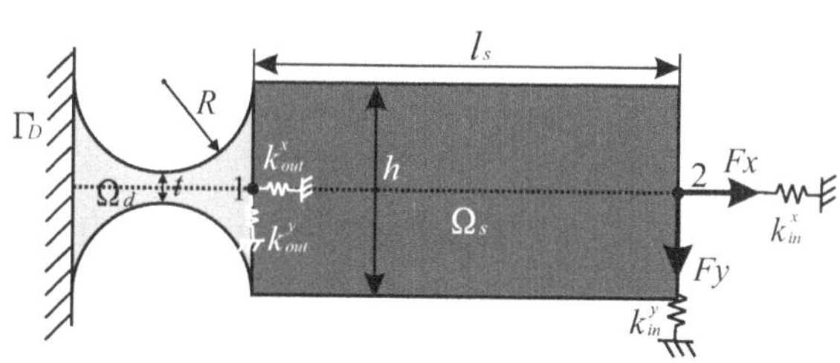

Fig. 2.21 Design domain and boundary condition of flexure hinges

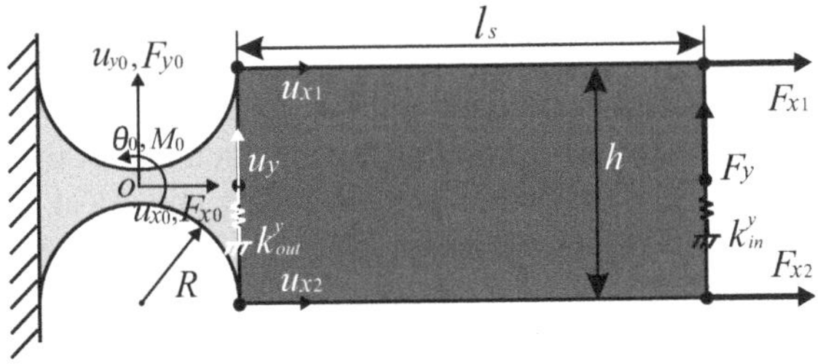

Fig. 2.22 Schematic representation of optimization model Θ_2

respectively. $\boldsymbol{\rho}$ is the vector of design variables, and ρ_i is the design variable of the ith element. $\boldsymbol{v}$ is an N-vector that contains all the element volumes, Vol_0 is the volume of the design domain, and V^* is the allowed volume fraction. To avoid singularity of the global stiffness matrix, ρ_{min} is typically set to 0.001. $\boldsymbol{K}$ is the global stiffness matrix, which can be written as

$$\boldsymbol{K} = \sum_{i=1}^{N} \rho_i{}^p \boldsymbol{K}_e + \boldsymbol{K}_s. \tag{2.50}$$

Note that the topology optimization modeling of a flexure hinge has other possibilities, for example, the loads can be directly applied to the right side of the design domain and distributed over several nodes to ensure the accuracy of FEA. Additionally, the optimization model also has other choices. One possible way is to apply three loads (F_{x0}, F_{y0} and M_0) and measure the three corresponding displacements (u_{x0}, u_{y0} and θ_0) at the center of the design domain (see Fig. 2.22). Then, the three displacements are optimized: the displacements u_{x0} and u_{y0} are minimized, and the angular displacement θ_0 is maximized. Here, we adopt the bilinear four-node square finite element, which has been widely used for topology optimization [3, 14, 30, 34, 39, 41]. Since each node has only two degrees of freedom, which are translations in the x- and y-directions, the moment M_0 cannot be directly applied at the node. However, this problem is easily overcome by applying loads F_{x1}, F_{x2} and F_y (see Fig. 2.22) at three points so that a moment is generated. Similarly, a rotation can be derived from the displacements u_{x1}, u_{x2} and u_y at three points.

An alternative optimization model Θ_2 is shown in Fig. 2.22. The loads F_{x0}, F_{y0} and M_0 can be equivalent to the form of the loads F_{x1}, F_{x2} and F_y, as follows

$$\begin{bmatrix} F_{x0} \\ F_{y0} \\ M_0 \end{bmatrix} = \begin{bmatrix} 1 & 1 & 0 \\ 0 & 0 & 1 \\ -h/2 & h/2 & l_s + R \end{bmatrix} \begin{bmatrix} F_{x1} \\ F_{x2} \\ F_y \end{bmatrix} \tag{2.51}$$

The displacement vectors $[u_{x1}, u_{x2}, u_y]^{\mathrm{T}}$ and $[u_{x0}, u_{y0}, \theta_0]^{\mathrm{T}}$ are connected by means of a matrix

$$\begin{bmatrix} u_{x1} \\ u_{x2} \\ u_y \end{bmatrix} = \begin{bmatrix} 1 & 0 & -h/2 \\ 1 & 0 & h/2 \\ 0 & 1 & R \end{bmatrix} \begin{bmatrix} u_{x0} \\ u_{y0} \\ \theta_0 \end{bmatrix} \tag{2.52}$$

Thus, the displacements u_{x0}, u_{y0} and θ_0 can be represented by the displacements u_{x1}, u_{x2} and u_y, as follows

$$\begin{bmatrix} u_{x0} \\ u_{y0} \\ \theta_0 \end{bmatrix} = \begin{bmatrix} 1/2 & 1/2 & 0 \\ R/h & -R/h & 1 \\ -1/h & 1/h & 0 \end{bmatrix} \begin{bmatrix} u_{x1} \\ u_{x2} \\ u_y \end{bmatrix} = \begin{bmatrix} \dfrac{u_{x1}+u_{x2}}{2} \\ u_y - \dfrac{R(u_{x2}-u_{x1})}{h} \\ \dfrac{u_{x2}-u_{x1}}{h} \end{bmatrix} \tag{2.53}$$

Therefore, a possible objective function for the optimization model Θ_2 can be stated as

$$\min: \quad -\omega_1\theta_0 + \omega_2 u_{y0} + (1-\omega_1-\omega_2)u_{x0}, \tag{2.54}$$

and the generate formulation for optimization model Θ_2 can be rewritten as

$$\begin{aligned} \Theta_2: \quad \min: \quad & \omega_2 u_y + \frac{1-\omega_1-\omega_2}{2}(u_{x1}+u_{x2}) - \frac{\omega_1+\omega_2 R}{h}(u_{x2}-u_{x1}) \\ \text{s.t}: \quad & \boldsymbol{Ku} = \boldsymbol{F} \\ & \frac{\boldsymbol{v}^T\boldsymbol{\rho}}{Vol_0} \le V^* \\ & 0 < \rho_{min} \le \rho_i \le 1, i = 1, 2 \ldots, N \end{aligned} \tag{2.55}$$

where ω_1 and ω_2 $(0 \le \omega_1, \omega_2, (\omega_1+\omega_2) \le 1)$ are weighting factors, $\boldsymbol{u}$ is the displacement field, and $\boldsymbol{F}$ is vector that contains forces F_{x1}, F_{x2} and F_y.

2.4.3 *Topology Results*

In this section, the numerical results for the removal of material from right-circular flexure hinges are presented. Bilinear four-node square elements are used in the FEA. The isotropic material has a Young's modulus of $E = 1$ and a Poison's ratio of $\mu = 0.3$. The forces F_x and F_y applied at the input port (point 2 in Fig. 2.21) are set to unity, and the input and output spring stiffnesses are $k_{in}^x = k_{out}^x = 1$ and $k_{in}^y = k_{out}^y = 1$. The length of the solid domain l_s is discretized into 300 finite elements, and the height of the design domain h is discretized into 100 finite elements. The sizes of t and R are measured by the number of finite elements. The available material volume constraint V^* will be stated for each result.

Tseytlin [38] categorized right circular flexure hinges into three groups depending on the t/R ratio: $t/R \le 0.07$, $0.07 < t/R < 0.2$, and $0.2 \le t/R \le 0.6$ are thin, intermediate and thick right-circular flexure hinges, respectively. Thus, topology optimization is performed for different t/R ratios that cover all three groups.

Figure 2.23 shows the final topology results of optimization model Θ_1 for several t/R ratios, namely, 0.05, 0.12, 0.22, 0.35 and 0.55, using the same parameters.

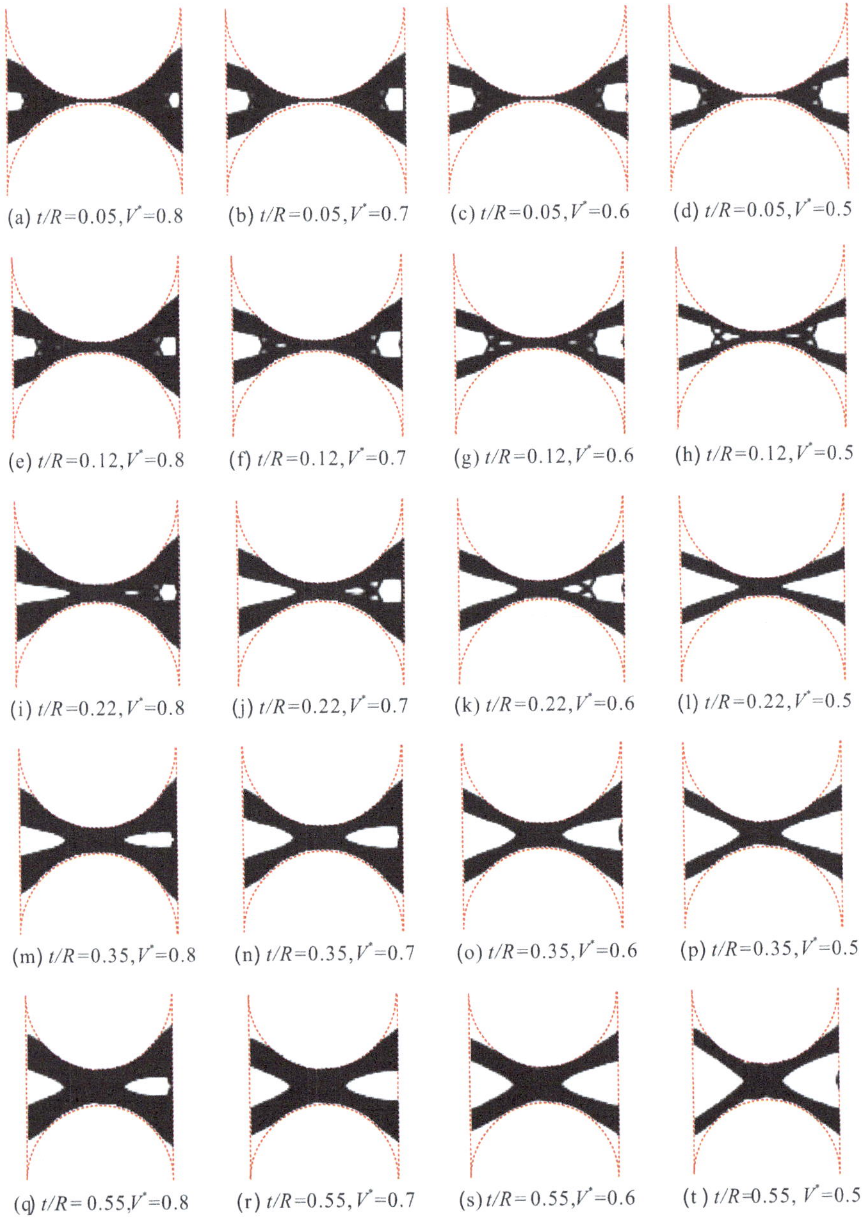

Fig. 2.23 Topology optimization results of optimization model Θ_1 for different t/R ratios and volume fractions

Table 2.2 The displacements of point 1 (see Fig. 2.21) in the x- and y-directions for all examples

t/R		$V^* = 1$	$V^* = 0.8$	$V^* = 0.7$	$V^* = 0.6$	$V^* = 0.5$
0.05	U_x	16.9351	17.5801	18.2733	19.2930	20.8536
	$U_y(\times 10^5)$	5.0634	5.0640	5.0650	5.0670	5.0710
0.12	U_x	7.9670	8.9737	10.1113	11.3488	13.3042
	$U_y(\times 10^4)$	2.8588	2.8748	3.2392	3.3975	3.7617
0.22	U_x	5.5314	6.6346	7.5898	8.6224	10.0749
	$U_y(\times 10^3)$	7.5284	9.2156	10.4400	11.8790	15.2270
0.35	U_x	4.3151	5.3416	6.1951	7.3340	9.0196
	$U_y(\times 10^3)$	3.2309	3.8879	4.5623	5.7712	8.6030
0.55	U_x	3.2563	4.3154	5.5641	6.8855	8.7839
	$U_y(\times 10^3)$	1.3181	1.6551	2.2848	2.9208	4.9118

The volume of removed material $(1 - V^*)$ is set to 20, 30, 40 and 50% of the design domain. The rough topologies are nearly the same for all examples, although the specific topology is slightly different for each one. Moreover, the four sharp corners of the design domain are removed in all the examples. The most significant change is that two notches are generated on either side of the design domain while the thin part is nearly unchanged. For $t/R = 0.05$ and 0.12, the shapes of the notches are similar to polygons. For $t/R = 0.22$, 0.35 and 0.55, the shapes of the notches are similar to a parabola. Additionally, the sizes of the parabolic notches increase with increasing volume of removed material in all examples.

Quantitative analysis is performed to compare the compliances before and after optimization. When the final topology result is obtained, FEA is performed to obtain the deformation of the flexure hinge. The loads F_x and F_y applied at the input port are also set to unity. Table 2.2 shows the displacements of point 1 (see Fig. 2.21) for all examples. U_x and U_y are the displacements of point 1 in the x- and y-directions due to forces F_x and F_y, respectively. $V^* = 1$ circular flexure hinges without material denotes the right-circular flexure hinges without material removal (unoptimized). The right-circular hinges without material removal have small displacements compared with the right-circular hinges with material removal. The displacements of point 1 increase with increasing volume of removed material in all examples, particularly when $t/R = 0.22, 0.35$ and 0.55. Figure 2.23 and Table 2.2 show that the appearance and size of parabolic notches have a major influence on the compliance of the multi-notched flexure hinges.

The topology optimization results obtained using optimization model Θ_2 for several examples are shown in Fig. 2.24 (the weighting factors ω_1 and ω_2 are set to $\frac{1}{3}$, the forces F_{x1}, F_{x2} and F_y are set to unity, and the spring stiffnesses are $k_{in}^y = k_{out}^y = 1$). The topology results are nearly equivalent to the corresponding topology results of optimization model Θ_1, which demonstrates the ability of topology optimization to identify new configurations of flexure hinges.

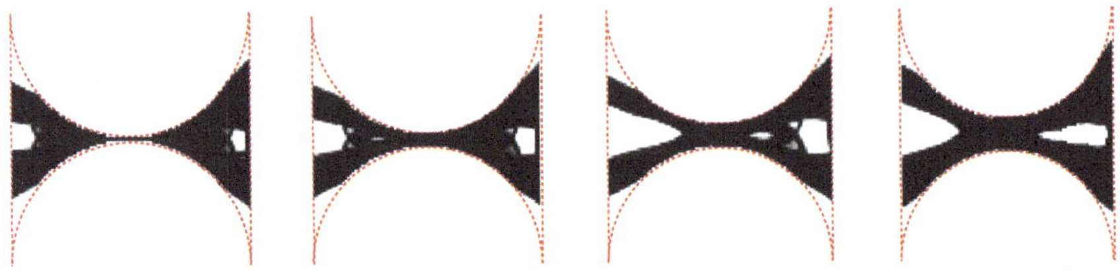

Fig. 2.24 Topology optimization results of optimization model Θ_2 for several examples

2.5 Post-design Modeling

A proposed second-stage design is developed in this section, which concentrates on modeling the flexure hinge. Based on the topology optimization results (see Fig. 2.11 and 2.23) of flexure hinges described in sections Sects. 2.2.5 and 2.4.3, we propose two new types of flexure hinges, that is, the quasi-V-shaped flexure hinge [19] and multi-notched flexure hinge [21].

2.5.1 Quasi-V-Shaped Flexure Hinge (QVFH)

Usually, a circle, ellipse, hyperbola or other common shape is selected as the profile of a flexure hinge, but these shapes may not be the best topologies for flexure hinges. Figure 2.11 shows that similar topologies are obtained and the shapes of the topologies are different from the shapes of the conventional hinges. The profile of the flexure hinge is formed by three different segments (as shown in Fig. 2.26). The middle segment is slightly different for different spring stiffness Ks, so we can obtain flexure hinges with varying performances by changing the thickness and width of the middle segment. Therefore, a new type of flexure hinges is designed based on the topology optimization results. Polynomial fitting is adopted to obtain the equations of the left and right segments of the flexure hinge's profile. It should be pointed out that the polynomial is not the only option for fitting the equations of flexure hinges; other functions can also be used. Generally, the fitting degree increases with increasing order, but higher order results in more complex equations for the compliance evaluation. Using the MATLAB curve fitting toolbox, we found that the coefficient of determination of a three-order polynomial is 0.9989, which indicates that it has high fitting degree. Furthermore, the three-order polynomial benefits the calculation. Therefore, the three-order polynomial is adopted to formulate the equations of the left and right segments. The middle segment of the profile is a straight line. The three-dimensional model of a QVFH is shown in Fig. 2.25, and the geometric parameters are illustrated in Fig. 2.26. The equation of a QVFH can be written as follows

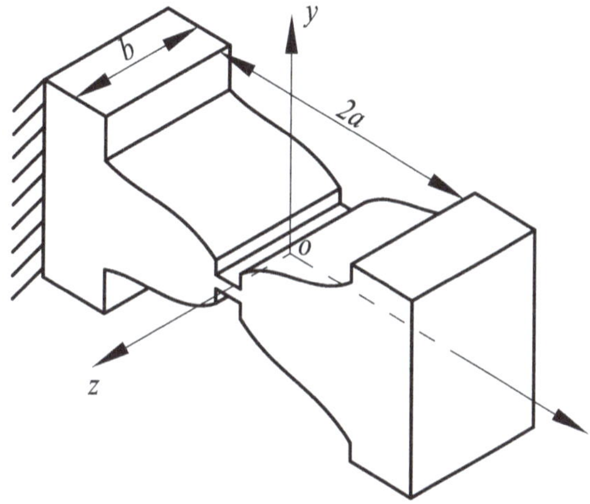

Fig. 2.25 Three-dimensional model of a flexure hinge

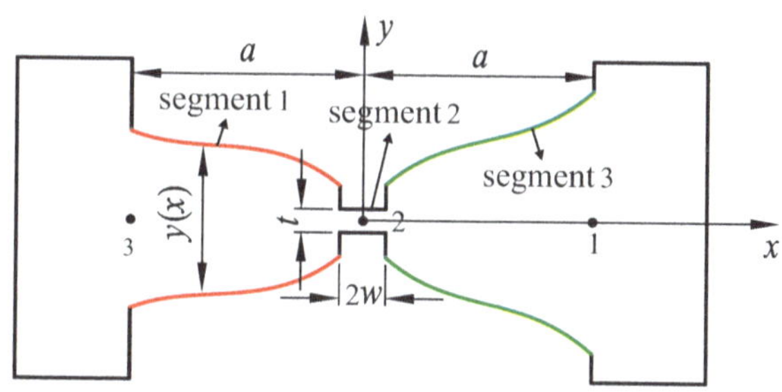

Fig. 2.26 Geometric parameters defining a flexure hinge

$$y(x) = \begin{cases} -\dfrac{1.666}{a^2}x^3 - \dfrac{3.218}{a}x^2 - 2.186x + 0.114a, & x \in [-a, -w) \\ t, & x \in [-w, w] \\ \dfrac{2.072}{a^2}x^3 - \dfrac{3.374}{a}x^2 + 2.302x + 0.108a, & x \in (w, a] \end{cases} \quad (2.56)$$

QVFHs are characterized in terms of compliance to external loading, accuracy of rotation, and the maximum stress. Castigliano's second theorem is used to solve the closed-form solutions. The theoretical analysis is performed in terms of two non-dimensional parameters, β and γ, which are defined as

$$\beta = \frac{t}{2w} and \qquad \gamma = \frac{t}{2a}. \quad (2.57)$$

2.5.1.1 Compliance Models

Compliance is one of the main parameters considered in flexure hinge design. The compliance equations of flexure hinges are derived by applying the basic theory of mechanics of materials. For a combination of loads, the principle of superposition can be used to calculate the total deformation since the deformations in all directions are sufficiently small.

Based on the previous assumptions, the hinge can be modeled as a cantilever beam. A schematic representation of a flexure hinge with loading is shown in Fig. 2.27. Point 1 is the free end with the load, point 2 is the center of rotation, and point 3 is the fixed end; these points correspond to positions 1, 2, and 3 in Fig. 2.26, respectively. By defining the load vector as

$$\mathbf{F} = [M_{z1}, F_{y1}, F_{x1}]^T \tag{2.58}$$

and the corresponding displacement vector as

$$\mathbf{X} = [\theta_1, y_1, x_1]^T, \tag{2.59}$$

the following displacement-load relationship at the free end is obtained [24]:

$$\mathbf{X} = \mathbf{CF}, \tag{2.60}$$

where $\mathbf{C}$ is the compliance matrix of the flexure hinge, which can be expressed as follows:

$$\mathbf{C} = \begin{bmatrix} C_{11} & C_{12} & 0 \\ C_{21} & C_{22} & 0 \\ 0 & 0 & C_{33} \end{bmatrix} \tag{2.61}$$

where $C_{12} = C_{21}$, according to the reciprocity principle. Each element in the compliance matrix is called a compliance factor.

Based on Castigliano's second theorem, the displacement-load relationship can be formulated as follows:

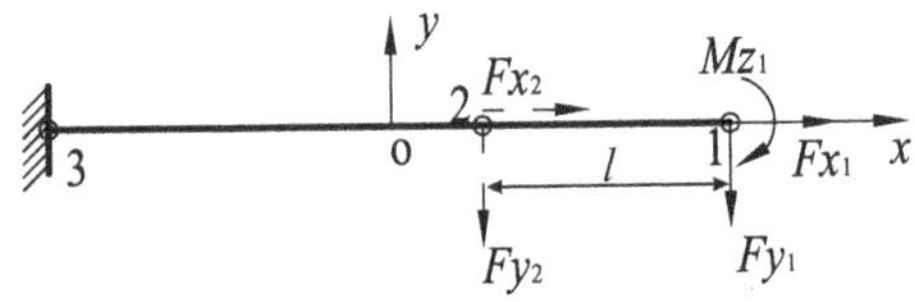

Fig. 2.27 Schematic representation of a flexure hinge with loading

$$\begin{cases} \theta_1 = \dfrac{\partial U_e}{\partial M_{z1}} \\ y_1 = \dfrac{\partial U_e}{\partial F_{y1}} \\ x_1 = \dfrac{\partial U_e}{\partial F_{x1}}. \end{cases} \tag{2.62}$$

Since the flexure hinge is subjected to bending and axial load, the elastic strain energy U_e comprises bending and axial terms and can be expressed as: [24]

$$U_e = \frac{1}{2}\int_{-a}^{a}\left\{\frac{[M_{z1} + (a-x)F_{y1}]^2}{EI(x)} + \frac{{F_{x1}}^2}{EA(x)}\right\}dx \tag{2.63}$$

where,

$$I(x) = \frac{b[y(x)]^3}{12}, \qquad A(x) = by(x). \tag{2.64}$$

The closed-form equations for the compliance factors can be derived by substituting Eq. (2.56) into Eq. (2.63) and combining with Eq. (2.62). Since the analytic solutions are too complex and prolix, adaptive Simpson integration is used to solve the integral. The three-order polynomial approximation is adopted to obtain the dimensionless equations of the flexure hinges, and the maximum error is less than 1% for all compliances and accuracies evaluations when compared with the analytic solution. Thus, the dimensionless compliance equations of the flexure hinges can be developed as follows:

$$C_{11} = \int_{-a}^{a}\frac{1}{EI(x)}dx = \frac{12}{bE}\int_{-a}^{a}\frac{1}{y(x)^3}dx = \frac{12}{bEt^2}f_1 \tag{2.65}$$

$$C_{12} = \int_{-a}^{a}\frac{a-x}{EI(x)}dx = \frac{12}{bE}\int_{-a}^{a}\frac{a-x}{y(x)^3}dx = \frac{12}{bEt}f_2 \tag{2.66}$$

$$C_{22} = \int_{-a}^{a}\frac{(a-x)^2}{EI(x)}dx = \frac{12}{bE}\int_{-a}^{a}\frac{(a-x)^2}{y(x)^3}dx = \frac{12}{bE}f_3 \tag{2.67}$$

$$C_{33} = \int_{-a}^{a}\frac{1}{EA(x)}dx = \frac{1}{bE}\int_{-a}^{a}\frac{1}{y(x)}dx = \frac{1}{bE}f_4 \tag{2.68}$$

where

$$\begin{aligned} f_1 = \Big(&0.004836 + 0.9777\beta - 0.5169\gamma + 0.02091\beta^2 + 3.523\beta\gamma \\ &-10.58\gamma^2 + 0.002515\beta^3 - 4.644\beta^2\gamma - 30.86\beta\gamma^2 + 309.4\gamma^3\Big)^{-1} \end{aligned} \tag{2.69}$$

$$f_2 = \left(0.00264 - 0.007062\beta - 0.3014\gamma + 0.004312\beta^2 + 2.784\beta\gamma + 5.587\gamma^2 + 0.0001465\beta^3 - 0.4447\beta^2\gamma - 11.78\beta\gamma^2 - 7.347\gamma^3\right)^{-1} \quad (2.70)$$

$$f_3 = \left(0.0004331 - 0.0009488\beta - 0.05639\gamma + 0.0004536\beta^2 + 0.1067\beta\gamma + 1.489\gamma^2 + 3.38 \times 10^{-5}\beta^3 - 0.04243\beta^2\gamma + 2.213\beta\gamma^2 - 9.826\gamma^3\right)^{-1} \quad (2.71)$$

$$f_4 = \left(0.07244 + 0.2896\beta + 1.006\gamma - 0.2404\beta^2 - 0.1132\beta\gamma - 8.002\gamma^2 + 0.0783\beta^3 - 0.1671\beta^2\gamma + 3.352\beta\gamma^2 + 33.43\gamma^3\right)^{-1} \quad (2.72)$$

2.5.1.2 Precision of Rotation

An ideal revolute joint should have a fixed center of rotation, but the center of rotation of flexure hinges will be different from the original position due to elastic deformation, which has a negative influence on the rotational accuracy of flexure hinges. When the drift of the center of rotation is large, the designed mechanism of flexure hinges will be unable to fulfil the requirements. Therefore, the accuracy of rotation is another important parameter for flexure hinge design. To analyze the properties of flexure hinges, the compliance of point 2 (see Fig. 2.26) is used to describe the accuracy of rotation.

The displacement-load relationship at the center of rotation can be expressed as

$$\begin{bmatrix} 0 \\ y_2 \\ x_2 \end{bmatrix} = \begin{bmatrix} 0 & 0 & 0 \\ C_{12}^{'} & C_{22}^{'} & 0 \\ 0 & 0 & C_{33}^{'} \end{bmatrix} \begin{bmatrix} M_{z1} \\ F_{y1} \\ F_{x1} \end{bmatrix} \quad (2.73)$$

where compliances $C_{12}^{'}$, $C_{22}^{'}$ and $C_{33}^{'}$ define the offset of the center of rotation and denote the accuracy of the rotation of flexure hinges.

The displacement of the center of rotation can be assessed using two fictitious loads, a horizontal load F_{x2} and a vertical load F_{y2} (as shown in Fig. 2.27), in addition to the actual load vector composed of M_{z1}, F_{y1} and F_{x1}. The displacements of the center of rotation are derived utilizing Castigliano's second theorem:

$$\begin{cases} y_2 = \dfrac{\partial U_e}{\partial F_{y2}} \\ x_2 = \dfrac{\partial U_e}{\partial F_{x2}} \end{cases} \tag{2.74}$$

Since the deviation of the center of rotation from the midpoint is relatively large for a transversely asymmetric flexure hinge [16], the assumed kinematical center is considered as the center of rotation of QVFHs.

When the center of rotation deviates from the midpoint, its position needs to be determined. A unique point is proposed as the center of rotation by Lipkin and Patterson [17]. It is identified from an investigation of compliant axes. The position of the center of rotation is defined as the distance to the free end (as shown in Fig. 2.27):

$$l = C_{12}/C_{11} \tag{2.75}$$

The elastic strain energy U_e consists of bending and axial terms:

$$U_e = \frac{1}{2}\int_{-a}^{a-l}\left\{\frac{[M_{z1} + (a-x)F_{y1} + (a-l-x)F_{y2}]^2}{EI(x)} + \frac{(F_{x1}+F_{x2})^2}{EA(x)}\right\}dx \tag{2.76}$$

The equations for accuracy of rotation is therefore as follows:

$$C_{12}^{'} = \int_{-a}^{a-l}\frac{a-l-x}{EI(x)}dx = \frac{12}{bE}\int_{-a}^{a-l}\frac{a-l-x}{y(x)^3}dx = \frac{12}{bEt}f_5 \tag{2.77}$$

$$C_{22}^{'} = \int_{-a}^{a-l}\frac{(a-l-x)(a-x)}{EI(x)}dx = \frac{12}{bE}\int_{-a}^{a-l}\frac{(a-l-x)(a-x)}{y(x)^3}dx = \frac{12}{bE}f_6 \tag{2.78}$$

$$C_{33}^{'} = \int_{-a}^{a-l}\frac{1}{EA(x)}dx = \frac{1}{bE}\int_{-a}^{a-l}\frac{1}{y(x)}dx = \frac{1}{bE}f_7 \tag{2.79}$$

where

$$f_5 = \left(-1.215 + 3.087\beta + 63.81\gamma + 7.551\beta^2 - 109.5\beta\gamma \right.$$
$$\left. -737.7\gamma^2 - 1.854\beta^3 - 142\beta^2\gamma + 3314\beta\gamma^2 - 1.304\times 10^4\gamma^3\right)^{-1} \tag{2.80}$$

$$f_6 = (-0.004165 - 0.03862\beta - 1.681\gamma + 0.07609\beta^2 + 16.2\beta\gamma$$
$$-45.32\gamma^2 - 0.01394\beta^3 - 5.183\beta^2\gamma - 116.7\beta\gamma^2 + 1057\gamma^3)^{-1} \tag{2.81}$$

$$f_7 = \left(0.1471 + 0.5418\beta + 1.781\gamma - 0.4497\beta^2 + 0.07881\beta\gamma - 13.16\gamma^2 + 0.1464\beta^3 - 0.3148\beta^2\gamma + 5.984\beta\gamma^2 + 51.58\gamma^3\right)^{-1} \quad (2.82)$$

2.5.1.3 Stress Considerations

Stress concentration is one of the main disadvantages of a flexure hinge and may result in fatigue failure. It is necessary to evaluate the maximum stress of flexure hinges when designing compliant mechanisms with flexure hinges. The stress levels at any cross section of planar flexure hinges consist of normal stresses and shear stresses. Since the shear effects are neglected, the normal stresses on flexure hinges are the result of bending and axial loading. The maximum stress produced by axial loading at the notch region can be calculated as [27]:

$$\sigma_{max,F_x} = k_a \frac{F_{x1}}{bt}, \quad (2.83)$$

where k_a is the stress concentration factor and can be found in reference [27].

For an applied bending, the maximum stress on the upper fiber at the notch region can be calculated as:

$$\sigma_{max,M_{z1}+F_{y1}} = k_b \frac{6(M_{z1} + F_{y1}a)}{bt^2}. \quad (2.84)$$

We use ANSYS to calculate $\sigma_{max,M_{z1}+F_{y1}}$ and then obtain the stress concentration factor k_b based on Eq. (2.84). The FEA results of k_b for several groups of flexure hinges are shown in Fig. 2.28. Two non-dimensional parameters, β and γ, are used to fit the FEA results, and a reduced-order polynomial expression of k_b is written as

$$k_b = \frac{0.5272}{\beta^4} - \frac{3.846\gamma + 3.221}{\beta^3} + \frac{18.04\gamma + 7.314}{\beta^2} - \frac{28.02\gamma + 7.369}{\beta} + 15.2\gamma + 3.83 \quad (2.85)$$

Considering that flexure hinges are mainly subjected to bending loads, the stress produced by the axial load is relatively small compared with the stress caused by bending. Therefore, the maximum stress caused by axial loading is neglected. Equations (2.84) and (2.85) can be used to evaluate the maximum stress of flexure hinges.

2.5.1.4 Finite Element and Experimental Verification

To verify the compliance equations of the QVFHs, FEA is conducted to evaluate the compliance and accuracy of rotation. Since the equations for calculating the stress level are obtained using FEA, the maximum stress does not have to be verified. ANSYS software is used to implement the FEA. The finite element model of a flexure

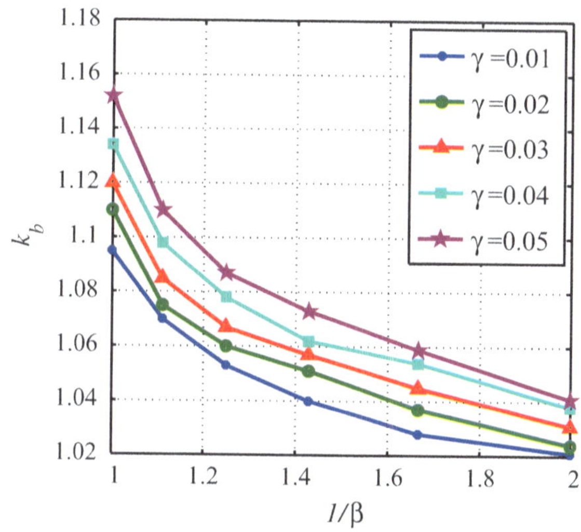

Fig. 2.28 FEA results for stress concentration factor k_b

hinge is generated using two-dimensional 8-node Plane 183 elements. There are two degrees of freedom on each node, which are translations in the x- and y-directions. This type of element is well suited to generate mesh for irregular shapes and curved boundaries. The mapping mesh method is employed to improve the computational accuracy. The finite element model of a flexure hinge is shown in Fig. 2.29.

All the flexure hinges have the same cross-sectional width $b = 10$ mm, and the material used for the flexure hinge is titanium alloy with a Young's modulus of 110 GPa and a Poisson's ratio of 0.33. The loads are $F_{x1} = 1$ N, $F_{y1} = 1$ N and $M_{z1} = 1$ Nmm.

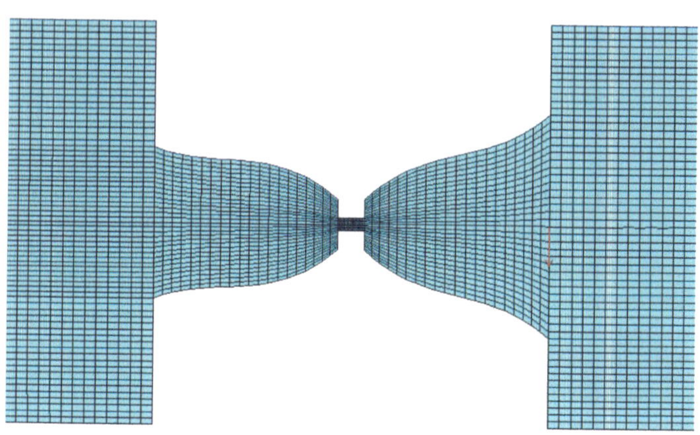

Fig. 2.29 Finite element model for a flexure hinge

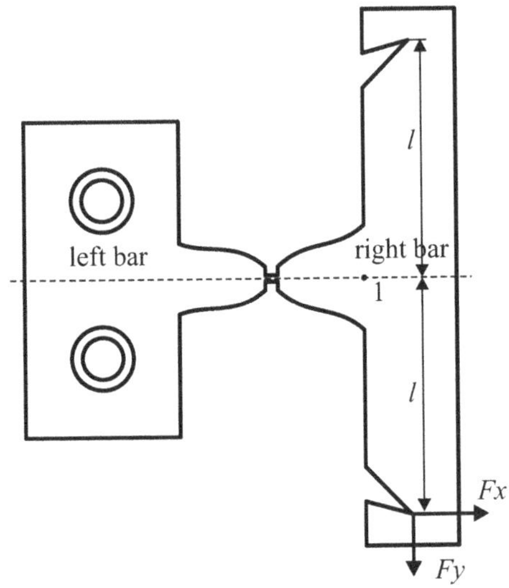

Fig. 2.30 Diagram of a sample profile

The FEA and analytical results of the compliance factors are shown in Table 2.3 for flexure hinges with different geometric parameters. Similarly, Table 2.4 shows the corresponding results of the accuracy factors.

As shown in Tables 2.3 and 2.4, the analytical and finite element results are in good agreement. The analytical model predictions are confirmed by the finite element simulations within 8% uncertainty.

The uncertainty increases when β approaches 1 or γ approaches 0.05. To reduce the uncertainty, β and γ should be less than 1 and 0.05, respectively. In addition, the results of the topology optimization in Fig. 2.11 indicate that the middle segment cannot be too long. If w is too large, the flexure hinge is similar to a simple beam. Therefore, we limit β to greater than 0.5. For the sake of comprehensive consideration, the limits of the two non-dimensional parameters are $0.5 \leq \beta \leq 1$ and $\gamma \leq 0.05$.

Experimentation is used to further verify the compliance factors C_{11}, C_{12} and C_{22}. The compliance factor C_{33} is not measured since it is very small. Two 10-mm-thick aluminum alloy flexure hinge specimens are machined by a wire electrical discharge machine with low wire traveling speed. The Young's modulus of the material is theoretically estimated to be $E = 71$ GPa with a Poisson's ratio $\mu = 0.33$. The other geometric dimensions are shown in Table 2.5. Each hinge is machined as an I-shaped specimen with the left and right bars separated by the hinge itself and a triangular cutout on each side of the right bar, as shown in Fig. 2.30.

The experimental setup comprises the micro-vision measurement system (0.28 μm resolution), computer, fixtures and flexure hinge specimens, as shown in Fig. 2.31. The micro-vision measurement system consists of an optical microscope with controllable zoom (magnification from 0.71× to 4.5×) and a CCD camera (JAI CV-A2, black and white) with a resolution of 1500 × 1200. A 10× objective, which

Table 2.3 Comparison of the compliance factors from the analytical results (denoted by An) and FEA results

Geometry			C_{11}(rad/Nm $\times 10^{-1}$)			C_{12}(N^{-1} $\times$ 10^{-3})			C_{22}(N^{-1} m $\times 10^{-5}$)			C_{33}(N^{-1} m $\times 10^{-9}$)		
a(mm)	t(mm)	w(mm)	An	FEA	Err%	An	FEA	Err%	An	FEA	Err%	An	FEA	Err%
5.0	0.2	0.2	5.536	5.572	0.65	2.800	2.786	0.50	1.390	1.392	0.14	4.973	4.988	0.30
5.0	0.4	0.2	0.763	0.832	8.29	0.383	0.416	7.93	0.195	0.210	7.69	4.070	4.231	3.81
5.0	0.4	0.4	1.416	1.428	0.84	0.710	0.714	0.56	0.358	0.361	0.83	4.660	4.684	0.51
5.0	0.5	0.5	0.918	0.952	3.57	0.461	0.476	3.15	0.234	0.241	2.90	4.533	4.619	1.86
10.0	0.2	0.2	5.483	5.340	2.67	5.480	5.340	2.62	5.480	5.360	2.24	5.182	5.266	1.59
10.0	0.4	0.2	0.710	0.760	6.58	0.711	0.763	6.82	0.715	0.760	5.92	4.270	4.381	2.53
10.0	0.4	0.4	1.384	1.373	0.80	1.380	1.372	0.58	1.390	1.380	0.72	4.973	4.989	0.32
10.0	0.5	0.5	0.891	0.893	0.22	0.891	0.893	0.22	0.895	0.899	0.44	4.890	4.910	0.41

Table 2.4 Comparison of the accuracy factors from the analytical results (denoted by An) and FEA results

Geometry			$C'_{12}(N^{-1} \times 10^{-5})$			$C'_{22}(N^{-1}m \times 10^{-7})$			$C'_{33}(N^{-1}m \times 10^{-9})$		
a(mm)	t(mm)	w(mm)	An	FEA	Err%	An	FEA	Err%	An	FEA	Err%
5.0	0.2	0.2	3.188	3.174	0.44	1.768	1.771	0.17	2.561	2.568	0.27
5.0	0.4	0.2	0.805	0.831	3.13	0.543	0.564	3.86	2.070	2.190	5.48
5.0	0.4	0.4	1.784	1.806	1.22	1.063	1.082	1.75	2.390	2.460	2.84
5.0	0.5	0.5	1.495	1.541	2.98	0.917	0.943	3.07	2.321	2.411	3.72
10.0	0.2	0.2	2.970	3.060	2.94	3.144	3.212	2.12	2.671	2.728	2.08
10.0	0.4	0.2	0.584	0.601	2.83	0.727	0.761	4.46	2.196	2.271	3.31
10.0	0.4	0.4	1.594	1.610	0.99	1.770	1.773	0.17	2.560	2.568	0.27
10.0	0.5	0.5	1.316	1.321	0.38	1.489	1.500	0.73	2.513	2.524	0.44

Table 2.5 Comparison of the compliance factors from the analytical results (denoted by An) and experimental results (denoted by Exp)

Geometry			C_{11}(rad/Nm $\times 10^{-2}$)			C_{12}($N^{-1} \times 10^{-4}$)			C_{22}(N^{-1} m $\times 10^{-6}$)		
a(mm)	t(mm)	w(mm)	An	Exp	Err%	An	Exp	Err%	An	Exp	Err%
15.00	1.02	1.04	3.62	3.75	3.62	5.43	5.62	3.49	8.20	8.55	4.27
15.02	1.51	1.23	1.39	1.52	9.35	2.09	2.28	9.09	3.20	3.41	6.56

has a depth of field of 13.5 μm, is used. The left bar of the flexure hinge is mounted horizontally on the platform to be rigidly held in place. The experimentat includes two setups to calculate the compliance factors C_{11}, C_{12} and C_{22}. In the first test (as shown in Fig. 2.31a), a transverse force F_x is applied to provide an equivalent moment $M_{z1} = F_x l$. The displacement of point 1 can be measured by the micro-vision measurement system. Knowing the moment and the displacement, the compliance factor C_{12} can be evaluated for each load step. By considering that the deformation is sufficiently small, the angular displacement of the flexure hinge can be written as $\theta_1 \approx \tan\theta_1$. Then, the compliance factor C_{11} can also be obtained from the test.

In the second test (as shown in Fig. 2.31b), a similar procedure is used to measure the displacement of point 1 under force F_y applied at the triangular cutout. Then, the displacement of point 1 under the force applied at position 1 can be obtained by an equivalent computation, so C_{22} can be obtained.

The flexure hinge specimens are loaded with seven different loads in each setup. The force is evaluated by placing known masses ranging from 0 to 350 g. The compliance factors are calculated as the average values. The analytical and experimental data are summarized in Table 2.5. The experimental results confirm the theoretical model predictions within 9% uncertainty.

The uncertainty is a result of several potential factors: (1) manufacturing uncertainty (the difference between the sample geometric parameters and the model geometric parameters); (2) measurement system uncertainty; and (3) loading uncertainty

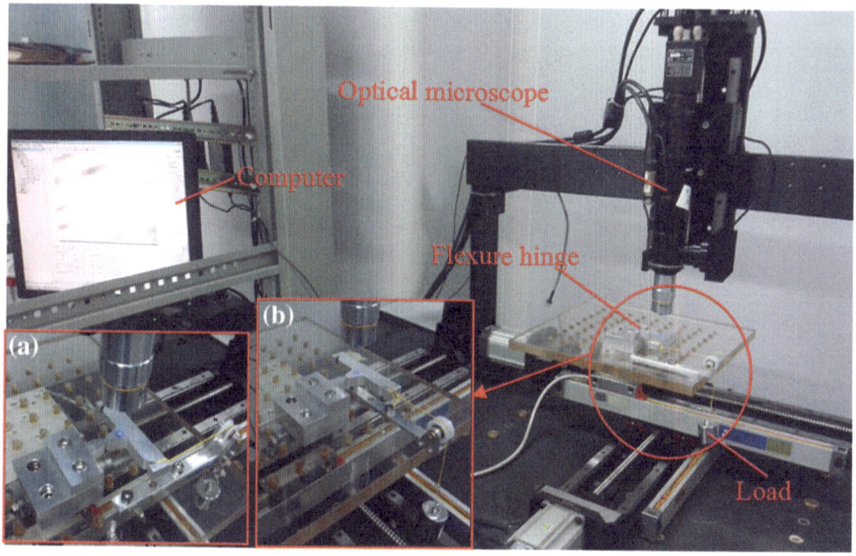

Fig. 2.31 The experimental setups: **a** setup for the evaluation of C_{11} and C_{12}; **b** setup for the evaluation of C_{22}

(the loads are determined by the masses, and uncertainty exists in the real and calibrated values of these masses).

2.5.1.5 Numerical Simulations

The established dimensionless equations can be used to investigate the influences of the material properties and geometric parameters on the compliance, accuracy of rotation and maximum stress. Numerical calculations are used to understand the characteristics of the QVFH. The material and geometric parameters of the flexure hinges are Young's modulus $E = 110\,\text{GPa}$, width $b = 10\,\text{mm}$ and minimum thickness $t = 0.5$ mm.

The numerical results for the compliances are shown in Fig. 2.32. Some conclusions can be drawn by analyzing the effect of the non-dimensional parameters β and γ on the compliance factors of the flexure hinges. All the compliance factors increase with decreasing β and γ in a nonlinear manner. Figure 2.32 shows that β has a larger influence than γ on compliance factors C_{11} and C_{33}, and γ has a larger influence than β on compliance factors C_{12} and C_{22}. Therefore, w plays a more important role than a in determining C_{11} and C_{33}, and a plays a more important role than w in determining C_{12} and C_{22}.

The rotational accuracies of the flexure hinges are shown in Fig. 2.33. All the accuracy factors increase with decreasing β and γ in a nonlinear manner. Figure 2.33a, b and c indicate that β has a larger influence than γ on $C_{33}^{'}$ and that γ has a larger

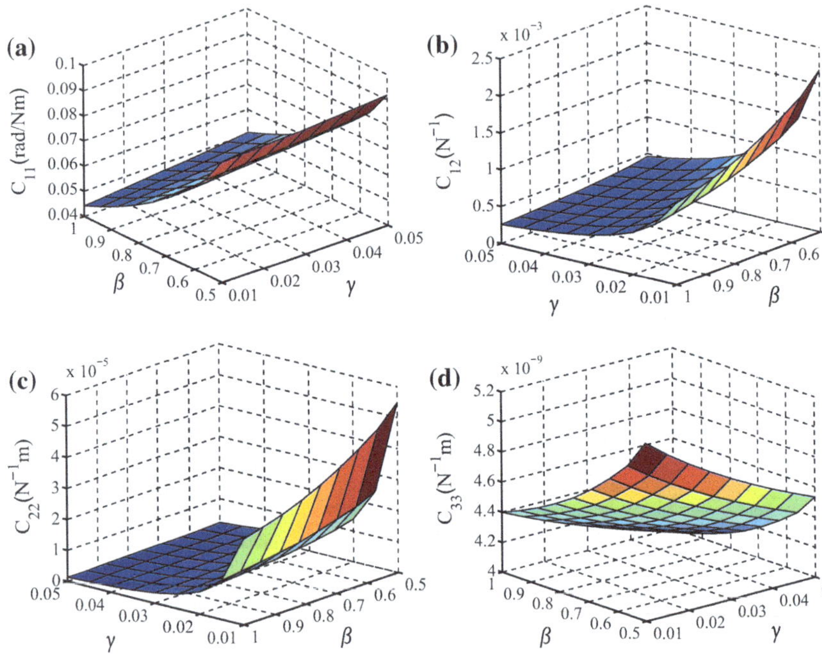

Fig. 2.32 Compliance factors of flexure hinges in terms of the non-dimensional parameters β and γ. **a** C_{11}, **b** C_{12}, **c** C_{22}, and **d** C_{33}

influence than β on C'_{12} and C'_{22}. Therefore, w plays a more important role than a in determining C'_{33}, and a plays a more important role than w in determining C'_{12} and C'_{22}.

2.5.2 Multi-notched Flexure Hinge

A multi-notched flexure hinge (MFH) consists of two right-circular and two parabolic notches. The geometric parameters and loads of the proposed flexure hinges are shown in Fig. 2.34, where B is the width, t denotes the distance between two circular notches, R is the radius, h denotes the height, and a is the distance from the apex of the parabola to the origin. The equations for the upper circular and left and right parabolic profiles of the proposed flexure hinges are denoted by $y_1(x)$, $y_2(x)$ and $y_3(x)$, respectively. The equations are given as follows:

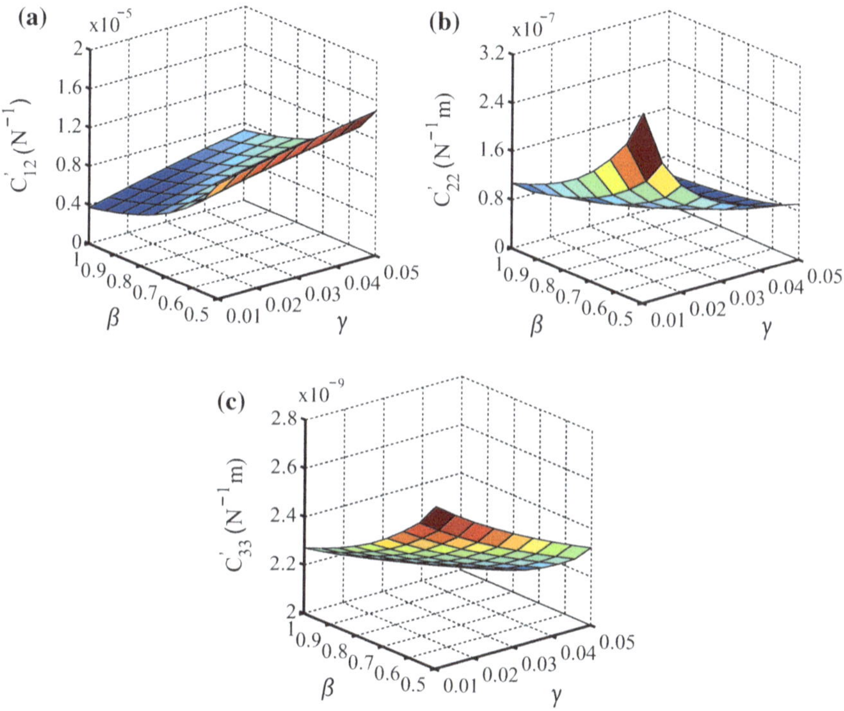

Fig. 2.33 Rotational accuracy factors of flexure hinges in terms of the non-dimensional parameters β and γ. **a** C'_{12}, **b** C'_{22}, and **c** C'_{33}

$$\begin{cases} y_1(x) = R + \dfrac{t}{2} - \sqrt{R^2 - x^2}, & x \in [-R, R] \\ y_2(x) = \sqrt{p\dfrac{t}{R}(-x-a)}, & x \in [-R, -a] \\ y_3(x) = \sqrt{p\dfrac{t}{R}(x-a)}, & x \in [a, R] \end{cases} \tag{2.86}$$

where p is used to control the size of the parabolic notch, which increases with increasing p/R. However, when p/R is too large, the parabolas will exceed the flexure hinges. Therefore, we provide an empirical limit of the p/R ratio ($0 \leq p/R \leq 0.5$). When $p/R = 0$, the proposed flexure hinges become right-circular flexure hinges.

As shown in Fig. 2.23, a is approximately equal to t when $t/R = 0.22$, and it is approximately half of t when $t/R = 0.55$. Thus, we assume that $a = t$ when $t/R = 0.2$ and that $a = t/2$ when $t/R = 0.6$. Additionally, it is assumed that a varies linearly with the t/R ratio as follows:

$$a = \frac{5}{4}t\left(1 - \frac{t}{R}\right). \tag{2.87}$$

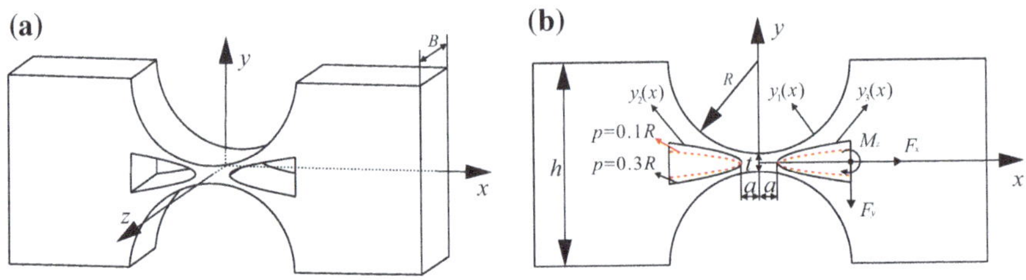

Fig. 2.34 Geometric parameters and loads of the multi-notched flexure hinges. (a) 3D plot of a flexure hinge, (b) 2D plot of a flexure hinge

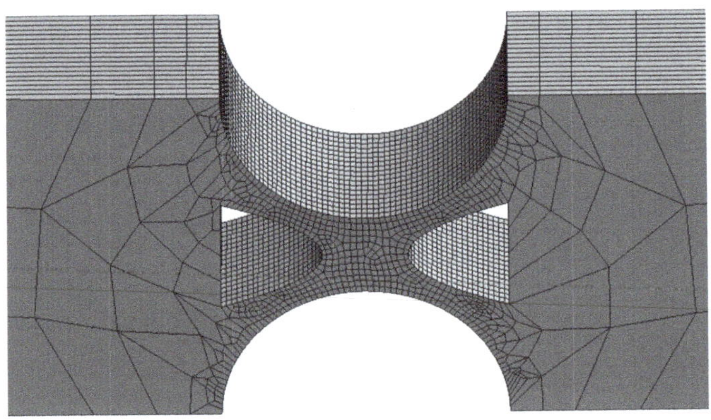

Fig. 2.35 Finite element model of a flexure hinge

2.5.2.1 Stiffness of Flexure Hinges

Analytical methods such as Castigliano's second theorem and empirical equations are often adopted to obtain the stiffness (reciprocal of the compliance) of flexure hinges. Since the errors in analytical methods are very large compare to those in the FEA results when the ratio t/R is large [43], and because the structure of the proposed flexure hinges is relatively complex and difficult to analyze, empirical equations through FEA are used to investigate the characteristics of the flexure hinges.

The dimensionless empirical equations for the flexure hinge are obtained using the finite element software ANSYS. For these calculations, the 3D flexure hinge models are generated using Solid186 elements. The elements are 20-node hexahedrons with three degrees of freedom at each node, which are translations in the nodal x-, y- and z-directions. This element type is well suited for modeling irregular shapes and curved boundaries without substantial loss of accuracy. The modeled flexure hinges have a width of 10 mm (structural steel), a Young's modulus of 200 GPa and a Poisson's ratio of 0.3. Figure 2.35 shows the finite element model of a multi-notched flexure hinge. Each flexure hinge model is fixed at the left end and loaded at the other end.

The empirical equations are formed based on the FEA results to estimate the stiffness in all directions for ranges of t/R $(0.2 \leq t/R \leq 0.6)$ and p/R $(0.1 \leq p/R \leq 0.5)$. When p/R is very small, the stress concentration is large near the apex, and the parabolic notch is difficult to process. Therefore, $0 \leq p/R < 0.1$ is not considered in this work. To mathematically characterize the flexure hinge, FEA models with various t/R and p/R ratios, which are set from 0.2 to 0.6 and from 0.1 to 0.5, respectively, with increments of 0.02, are generated in ANSYS. Unit loads F_x, F_y and M_z are applied on each model, and the corresponding deformations are read. Fifth-order polynomial functions are chosen to obtain the dimensionless empirical equations. The radius R in the finite element calculations is 10 mm. Thus, the dimensionless stiffness equations can be obtained as follows:

$$\begin{aligned} K_{\alpha_z M_z} &= \frac{12}{E \times B \times t^2}\left(\frac{1}{C_{11}}\right) \\ &= -0.07076 - 0.9568\beta + 3.298\gamma - 0.01423\beta^2 + 7.481\beta\gamma \\ &\quad - 13.47\gamma^2 - 2.7\beta^2\gamma - 29.15\beta\gamma^2 + 35.88\gamma^3 + 12.84\beta^2\gamma^2 \\ &\quad + 41.6\beta\gamma^3 - 48.2\gamma^4 - 13.38\beta^2\gamma^3 - 19.74\beta\gamma^4 + 24.48\gamma^5 \end{aligned} \tag{2.88}$$

$$\begin{aligned} K_{\alpha_z F_y} &= \frac{12}{E \times B \times t}\left(\frac{1}{C_{12}}\right) = K_{\Delta_y M_z} = \frac{12}{E \times B \times t}\left(\frac{1}{C_{21}}\right) \\ &= -0.03154 - 0.168\beta + 0.5316\gamma + 0.06073\beta^2 + 1.75\beta\gamma \\ &\quad - 1.577\gamma^2 - 1.31\beta^2\gamma - 8.084\beta\gamma^2 + 7.144\gamma^3 + 5.08\beta^2\gamma^2 \\ &\quad + 11.17\beta\gamma^3 - 11.23\gamma^4 - 4.848\beta^2\gamma^3 - 5.302\beta\gamma^4 + 5.932\gamma^5 \end{aligned} \tag{2.89}$$

$$\begin{aligned} K_{\Delta_y F_y} &= \frac{12}{E \times B}\left(\frac{1}{C_{22}}\right) \\ &= -0.00638 - 0.06666\beta + 0.09191\gamma + 0.1012\beta^2 + 0.7627\beta\gamma \\ &\quad - 0.5035\gamma^2 - 1.151\beta^2\gamma - 3.019\beta\gamma^2 + 2.993\gamma^3 + 3.695\beta^2\gamma^2 \\ &\quad + 3.235\beta\gamma^3 - 3.798\gamma^4 - 3.4\beta^2\gamma^3 - 1.166\beta\gamma^4 + 1.609\gamma^5 \end{aligned} \tag{2.90}$$

$$\begin{aligned} K_{\Delta_x F_x} &= \frac{1}{E \times B}\left(\frac{1}{C_{33}}\right) \\ &= 0.01101 - 0.1587\beta + 1.064\gamma - 0.0115\beta^2 - 0.3126\beta\gamma \\ &\quad - 2.503\gamma^2 + 0.8907\beta^2\gamma - 1.568\beta\gamma^2 + 5.425\gamma^3 - 0.203\beta^2\gamma^2 \\ &\quad + 1.931\beta\gamma^3 - 6.358\gamma^4 + 0.1344\beta^2\gamma^3 - 0.9624\beta\gamma^4 + 2.897\gamma^5 \end{aligned} \tag{2.91}$$

where $\beta = p/R$, $\gamma = t/R$, and $K_{\alpha_z M_z}$ and $K_{\alpha_z F_y}$ are the dimensionless rotation stiffnesses about the z-axis for moment M_z and force F_y, respectively. $K_{\Delta_y M_z}$ and $K_{\Delta_y F_y}$ are the dimensionless stiffnesses in the y-direction due to moment M_z and force F_y, respectively. $K_{\Delta_x F_x}$ is the dimensionless stiffness in the x-direction due to force F_x.

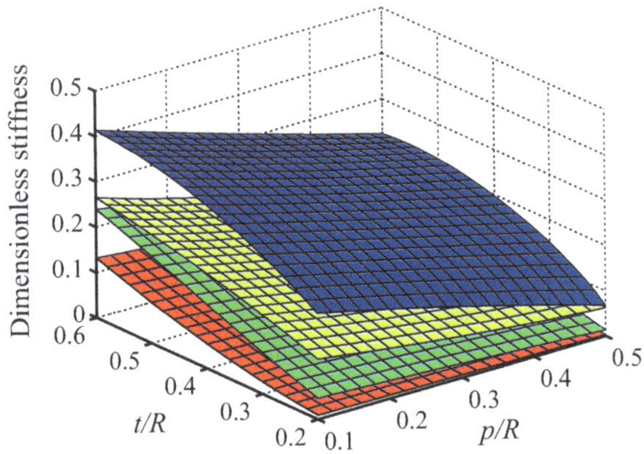

Fig. 2.36 Three-dimensional plot of the dimensionless stiffnesses. (Blue denotes $K_{\alpha_z M_z}$, green denotes $K_{\alpha_z F_y}$ and $K_{\Delta_y M_z}$, red denotes $K_{\Delta_y F_y}$, and yellow denotes $K_{\Delta_x F_x}$.)

For Eqs. (2.88)–(2.91), the correlation coefficients are 0.9979, 0.9997, 0.9989 and 0.9993, respectively. The dimensionless stiffnesses of the flexure hinges are also graphically illustrated in Fig. 2.36, where all the dimensionless stiffnesses increase non-linearly with increasing t/R and decrease non-linearly with increasing p/R.

2.5.2.2 Precision of Rotation

When loads are applied at the free end of the flexure hinge, the moment M_z and force F_y make the rotational center translate in the y-direction, and the force F_x makes the rotational center translate in the x-direction. Therefore, the rotational center of the flexure hinge shifts from its original position, which is not beneficial for the motion precision. If the drift of the rotational center becomes excessive, the flexure-based compliant mechanisms will be unable to fulfil the requirements. Thus, high rotational precision is a fundamental requirement for flexure hinges. To perform quantitative analysis, the stiffness of the rotational center (the geometric symmetry center of the flexure hinge) is typically used to denote the rotational precision [24, 36]. A larger stiffness of the rotational center corresponds to a higher rotational precision of the flexure hinge. The dimensionless stiffness equations for the rotational center are given as follows:

$$\begin{aligned} K_{\delta_y M_z} &= \frac{1}{E \times B \times t}\left(\frac{1}{C'_{21}}\right) \\ &= -0.08857 + 0.438\beta + 1.501\gamma - 0.1649\beta^2 - 5.526\beta\gamma \\ &\quad - 4.862\gamma^2 + 1.847\beta^2\gamma + 22.51\beta\gamma^2 + 8.461\gamma^3 - 7.623\beta^2\gamma^2 \\ &\quad - 36.61\beta\gamma^3 - 6.704\gamma^4 + 8.31\beta^2\gamma^3 + 21.7\beta\gamma^4 + 1.766\gamma^5 \end{aligned} \tag{2.92}$$

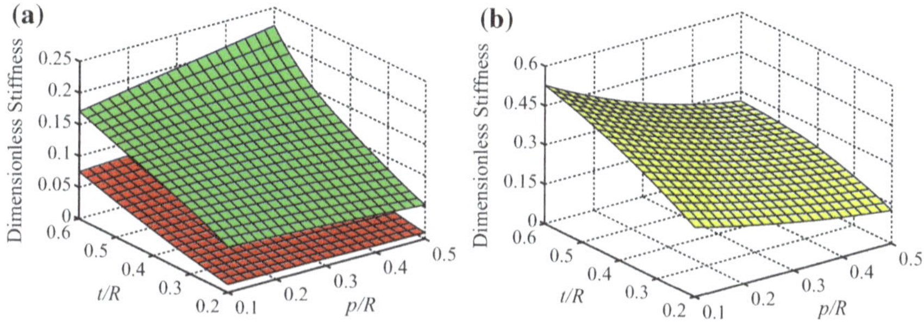

Fig. 2.37 Three-dimensional plots of the dimensionless stiffnesses of the rotational center. (a) dimensionless stiffnesses $K_{\delta_y M_z}$ and $K_{\delta_y F_y}$, (b) dimensionless stiffness $K_{\delta_x F_x}$. (Green denotes $K_{\delta_y M_z}$, red denotes $K_{\delta_y F_y}$, and yellow denotes $K_{\delta_x F_x}$.)

$$\begin{aligned} K_{\delta_y F_y} &= \frac{1}{E \times B}\left(\frac{1}{C'_{22}}\right) \\ &= -0.02544 + 0.1007\beta + 0.3161\gamma - 0.0299\beta^2 - 1.237\beta\gamma \\ &\quad - 0.9886\gamma^2 + 0.4024\beta^2\gamma + 4.979\beta\gamma^2 + 2.32\gamma^3 - 1.785\beta^2\gamma^2 \\ &\quad - 8.083\beta\gamma^3 - 2.422\gamma^4 + 1.903\beta^2\gamma^3 + 4.936\beta\gamma^4 + 0.9616\gamma^5 \end{aligned} \tag{2.93}$$

$$\begin{aligned} K_{\delta_x F_x} &= \frac{1}{E \times B}\left(\frac{1}{C'_{33}}\right) \\ &= 0.01453 - 0.3042\beta + 2.219\gamma - 0.0125\beta^2 - 0.806\beta\gamma \\ &\quad - 5.402\gamma^2 + 1.723\beta^2\gamma - 2.302\beta\gamma^2 + 11.6\gamma^3 - 0.389\beta^2\gamma^2 \\ &\quad + 2.413\beta\gamma^3 - 13.33\gamma^4 + 0.3683\beta^2\gamma^3 - 1.103\beta\gamma^4 + 5.954\gamma^5 \end{aligned} \tag{2.94}$$

where $K_{\delta_y M_z}$ and $K_{\delta_y F_y}$ are the dimensionless stiffnesses of the rotational center in the y-direction due to moment M_z and force F_y, respectively. $K_{\delta_x F_x}$ is the dimensionless stiffness of the rotational center in the x-direction due to force F_x.

The correlation coefficients of Eqs. (2.92)–(2.94) are 0.9998, 0.9999 and 0.9993, respectively. The dimensionless stiffnesses of the rotational center are graphically illustrated in Fig. 2.37. The dimensionless stiffnesses $K_{\delta_y M_z}$ and $K_{\delta_y F_y}$ (see Fig. 2.37a) increase non-linearly with increasing t/R, whereas they have different variations with increasing p/R. When t/R is small, $K_{\delta_y M_z}$ and $K_{\delta_y F_y}$ decrease with increasing p/R. When t/R is large, $K_{\delta_y M_z}$ and $K_{\delta_y F_y}$ increase as p/R increases. For dimensionless stiffness $K_{\delta_x F_x}$ (see Fig. 2.37b), $K_{\delta_x F_x}$ increases non-linearly with increasing t/R and decreases non-linearly with increasing p/R.

2.5.2.3 Stress Considerations

Considering that flexure hinges are mainly subjected to bending and axial loads, dimensionless empirical equations for evaluating the maximum stresses of flexure hinges subject to pure moment M_z and forces F_x and F_y are provided as follows:

$$\begin{aligned} S_{\sigma_{M_z}} &= \frac{\sigma_{M_z} \times B \times t^2}{6 \times M_z} \\ &= 9.093 - 2.712\beta - 94.83\gamma + 57.13\beta^2 - 82.3\beta\gamma \\ &\quad + 482.5\gamma^2 - 296.3\beta^2\gamma + 540.6\beta\gamma^2 - 1257\gamma^3 + 526.2\beta^2\gamma^2 \\ &\quad - 1072\beta\gamma^3 + 1619\gamma^4 - 297.7\beta^2\gamma^3 + 683.7\beta\gamma^4 - 810.7\gamma^5 \end{aligned} \tag{2.95}$$

$$\begin{aligned} S_{\sigma_{F_x}} &= \frac{\sigma_{F_x} \times B \times t}{F_x} \\ &= 2.067 - 2.421\beta + 2.129\gamma + 34.06\beta^2 - 32.1\beta\gamma \\ &\quad - 45.19\gamma^2 - 179.5\beta^2\gamma + 269.6\beta\gamma^2 + 133.1\gamma^3 + 317.6\beta^2\gamma^2 \\ &\quad - 567.1\beta\gamma^3 - 141.6\gamma^4 - 177\beta^2\gamma^3 + 374.9\beta\gamma^4 + 47.48\gamma^5 \end{aligned} \tag{2.96}$$

$$\begin{aligned} S_{\sigma_{F_y}} &= \frac{\sigma_{F_y} \times B \times t}{12 \times F_y} \\ &= 31.2 - 15.71\beta - 324.5\gamma + 205.8\beta^2 - 174.9\beta\gamma \\ &\quad + 1556\gamma^2 - 1272\beta^2\gamma + 1552\beta\gamma^2 - 3950\gamma^3 + 2639\beta^2\gamma^2 \\ &\quad - 3555\beta\gamma^3 + 5090\gamma^4 - 1802\beta^2\gamma^3 + 2546\beta\gamma^4 - 2603\gamma^5 \end{aligned} \tag{2.97}$$

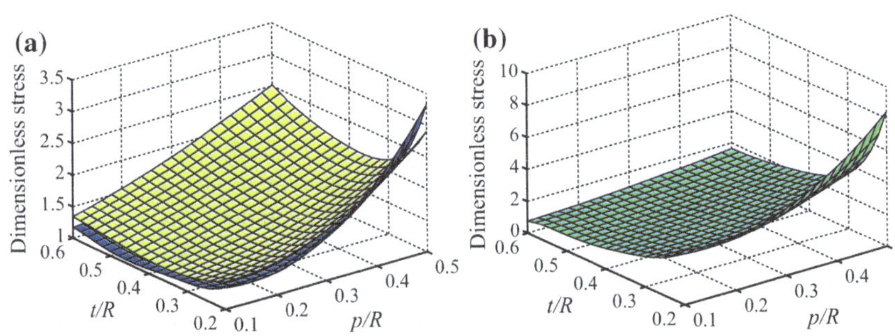

Fig. 2.38 Three-dimensional plot of the dimensionless stresses. **a** Dimensionless stresses $S_{\sigma_{M_z}}$ and $S_{\sigma_{F_x}}$; **b** dimensionless stress $S_{\sigma_{F_y}}$. (Blue denotes $S_{\sigma_{M_z}}$, yellow denotes $S_{\sigma_{F_x}}$, and green denotes $S_{\sigma_{F_y}}$)

where $S_{\sigma_{M_z}}$ is the dimensionless rotation stress due to moment M_z, $S_{\sigma_{F_x}}$ is the dimensionless stress due to axial force F_x, and $S_{\sigma_{F_y}}$ is the dimensionless stress due to force F_y. σ_{M_z}, σ_{F_x} and σ_{F_y} are the maximum stresses due to M_z, F_x and F_y, respectively.

The correlation coefficients of Eqs. (2.95)–(2.97) are 0.9725, 0.9881 and 0.9891, respectively. Figure 2.38 shows the dimensionless stress of the flexure hinges. The dimensionless stresses $S_{\sigma_{M_z}}$ and $S_{\sigma_{F_x}}$ (see Fig. 2.38a) increase non-linearly with increasing p/R and decrease first and then increase with increasing t/R. Additionally, dimensionless stress $S_{\sigma_{F_y}}$ (see Fig. 2.38b) decreases non-linearly with increasing t/R and increases non-linearly with increasing p/R.

2.5.2.4 Comparison with Right-Circular Flexure Hinges

The MFHs are compared with right-circular flexure hinges in terms of dimensionless stiffness, rotational precision and maximum stress.

Many studies in the literature address the stiffness equations of right-circular flexure hinges, and the accuracies of the stiffness equations are different than those of the FEA results. Among these studies, the equations proposed in reference [29] for the rotational stiffness and stress have the highest accuracy. Those equations are

$$\begin{aligned} K_{\alpha_z M_z R} &= \frac{12}{E \times B \times t^2}\left(\frac{1}{C_{11}}\right) \\ &= -0.0089 + 1.3556\sqrt{\frac{t}{2R}} - 0.5227\left(\sqrt{\frac{t}{2R}}\right)^2 \end{aligned} \tag{2.98}$$

$$S_{\sigma_{\alpha_z R}} = \frac{\sigma_{\alpha_z}}{E \times \alpha_z} = -0.0028 + 0.6397\sqrt{\frac{t}{2R}} - 0.0856\left(\sqrt{\frac{t}{2R}}\right)^2 \tag{2.99}$$

The equations proposed in reference [43] for the stiffnesses in the x- and y-directions have the highest accuracy. Those equations are

$$\begin{aligned} K_{\Delta_y F_y R} &= \frac{1}{E \times B}\left(\frac{1}{C_{22}}\right) \\ &= -0.018781\left(\frac{t}{R}\right)^6 + 0.062278\left(\frac{t}{R}\right)^5 - 0.088075\left(\frac{t}{R}\right)^4 \\ &+ 0.064783\left(\frac{t}{R}\right)^3 + 0.021734\left(\frac{t}{R}\right)^2 - 0.00083463\left(\frac{t}{R}\right) + 1.92 \times 10^{-5} \end{aligned} \tag{2.100}$$

$$\begin{aligned} K_{\Delta_x F_x R} &= \frac{1}{E \times B}\left(\frac{1}{C_{33}}\right) \\ &= 1.2031\left(\frac{t}{R}\right)^5 - 3.0831\left(\frac{t}{R}\right)^4 + 3.1152\left(\frac{t}{R}\right)^3 \\ &\quad - 1.5469\left(\frac{t}{R}\right)^2 + 0.98683\left(\frac{t}{R}\right) + 0.036343 \end{aligned} \tag{2.101}$$

However, there are no accurate design equations for estimating the rotational precision. Therefore, dimensionless empirical equations for the rotational precision of right-circular hinges are developed based on the FEA results in this work.

$$\begin{aligned} K_{\delta_y M_z R} &= \frac{1}{E \times B \times t}\left(\frac{1}{A_{21}}\right) \\ &= -0.1847\left(\frac{t}{R}\right)^5 - 0.06319\left(\frac{t}{R}\right)^4 + 0.7369\left(\frac{t}{R}\right)^3 \\ &\quad - 0.8826\left(\frac{t}{R}\right)^2 + 0.5472\left(\frac{t}{R}\right) - 0.008579 \end{aligned} \tag{2.102}$$

$$\begin{aligned} K_{\delta_y F_y R} &= \frac{1}{E \times B}\left(\frac{1}{A_{22}}\right) \\ &= -0.3946\left(\frac{t}{R}\right)^5 + 0.9347\left(\frac{t}{R}\right)^4 - 0.8779\left(\frac{t}{R}\right)^3 \\ &\quad + 0.4422\left(\frac{t}{R}\right)^2 + 0.008854\left(\frac{t}{R}\right) - 0.001045 \end{aligned} \tag{2.103}$$

Note that multi-notched flexure hinges have two non-dimensional parameters, t/R and p/R, whereas right-circular flexure hinges have only one parameter, t/R. Furthermore, note that the stiffnesses and stresses of right-circular flexure hinges are constant when p/R changes. For comparison, the dimensionless stiffnesses and stresses of the two flexure hinges are shown in Fig. 2.39, and the relative stiffnesses and stresses of the two hinges are also plotted. The z-axis in Fig. 2.39 adopts a non-uniform scale, and the subscript ($_R$) indicates a right-circular flexure hinge.

Based on the established dimensionless empirical equations, the comparisons of the stiffnesses and relative stiffnesses of the two flexure hinges are shown in Fig. 2.39a–c.The right-circular flexure hinges are stiffer than the multi-notched flexure hinges, and the relative stiffness increases with increasing p/R. This result indicates that the multi-notched flexure hinges have larger compliance than the right-circular flexure hinges. The bending compliance of the multi-notched flexure hinge is approximately 5 times larger than that of the right-circular hinge when $t/R = 0.2$ and $p/R = 0.5$. Moreover, the axial compliance of the multi-notched hinge is larger than that of the right-circular hinge (up to approximately 3 times when $p/R = 0.5$).

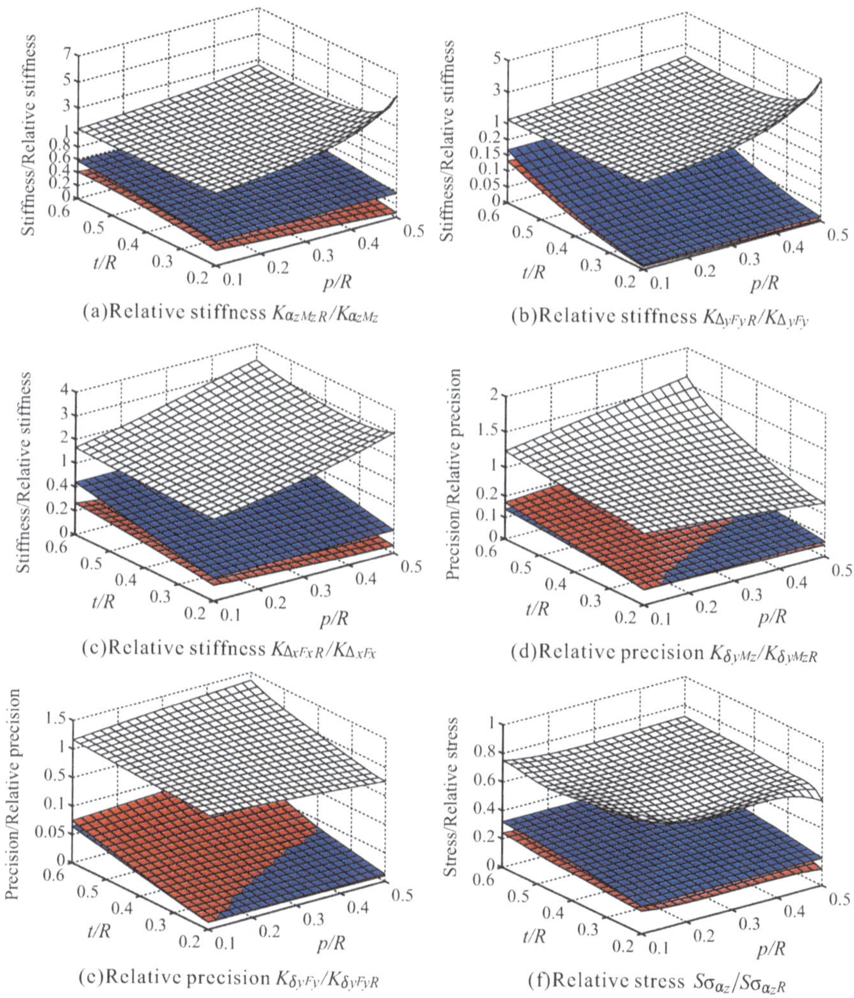

(a) Relative stiffness $K_{\alpha_z M_z R}/K_{\alpha_z M_z}$
(b) Relative stiffness $K_{\Delta_y F_y R}/K_{\Delta_y F_y}$
(c) Relative stiffness $K_{\Delta_x F_x R}/K_{\Delta_x F_x}$
(d) Relative precision $K_{\delta_y M_z}/K_{\delta_y M_z R}$
(e) Relative precision $K_{\delta_y F_y}/K_{\delta_y F_y R}$
(f) Relative stress $S_{\sigma_{\alpha_z}}/S_{\sigma_{\alpha_z R}}$

Fig. 2.39 Comparison of the multi-notched and right-circular flexure hinges. (Blue denotes right-circular flexure hinges (subscript ($_R$)), red denotes multi-notched flexure hinges, white denotes the comparative value)

The comparison of the rotational precision of the two hinges is shown in Fig. 2.39d and e. The rotational precision of the multi-notched flexure hinges is lower than that of the right-circular flexure hinges when t/R is small and p/R is large. With increasing t/R, the rotational precision of the multi-notched flexure hinges increases significantly and exceeds that of the right circular flexure hinges. The rotational precision of the multi-notched hinge is approximately 1.5 times greater than that of the right-circular hinge when $t/R = 0.6$ and $p/R = 0.5$.

To compare the maximum stress of the two flexure hinges, we assume that the flexure hinges are only subjected to bending M_z and have equal rotation angles α_z. Equations (2.88) and (2.95) together can be used to evaluate the dimensionless rotation stress of the multi-notched flexure hinges in terms of rotation angle α_z:

$$S_{\sigma_{\alpha_z}} = \frac{\sigma_{\alpha_z}}{E \times \alpha_z} = \frac{1}{2} \times \frac{\sigma_{M_z} \times B \times t^2}{6 \times M_z} \times \frac{12}{E \times B \times t^2} \left(\frac{1}{C_{11}} \right), \tag{2.104}$$

where σ_{α_z} is the maximum stress due to the rotation angle α_z.

Figure 2.39f compares the dimensionless rotation stress for the two flexure hinges. As shown in this figure, the multi-notched flexure hinges have lower rotation stress than that of the right-circular flexure hinges (down to approximately 0.7 times).

2.6 Conclusions

This chapter presented a method for the topology optimization of flexure hinges based on continuum-type topology optimization techniques. Basic optimization models for the topology optimization of translational and revolute hinges are developed. Numerical results are used to illustrate the validity of the proposed method. The numerical results confirm that for translational hinges, the existing designs are already optimal at the topology level, but the topologies of revolute flexure hinges are very different from the topologies of conventional flexure hinges.

To limit the maximum stress of the flexure hinge, a global stress constraint is added to the hinge topology optimization. The optimization model for the stress-constrained topology optimization of a flexure hinge is established, and the sensitivity analysis of the objective and constraint functions is derived. When the stress constraint is added to the flexure hinge optimization problem, it has a strong influence on the final topology results. The sharp corners are eliminated, and the profile of the middle segment becomes smooth. As the stress limit is lowered, the stress distributions transition from being local to being distributed, and the flexure hinge gradually tends towards a corner-filleted flexure hinge.

Some material is removed from the hinge using a topology optimization approach to improve the performances of the conventional flexure hinges. Optimization models are developed to remove material from the flexure hinge. The final topology results show that two notches are generated on either side of the flexure hinge while the thin part is nearly unchanged.

Based on the topology optimization results of flexure hinges, two new types of flexure hinges, i.e., quasi V-shaped and multi-notched flexure hinges, are designed and analyzed. To characterize the performances of the two flexure hinges in terms of their compliance, precision of rotation, and maximum stress levels, the dimensionless equations for the two flexure hinges are derived using Castiglianos second theorem and the finite element method.

References

1. Awtar, S., Slocum, A.H., Sevincer, E.: Characteristics of beam-based flexure modules. J. Mech. Des. **129**(6), 625–639 (2007)
2. Barthelemy, J.F., Haftka, R.T.: Approximation concepts for optimum structural designła review. Structural Optimization **5**(3), 129–144 (1993)
3. Bendsøe, M.P.: Optimal shape design as a material distribution problem. Structural Optimization **1**(4), 193–202 (1989)
4. Bendsoe, M.P., Sigmund, O.: Topology Optimization: Theory, Methods, and Applications. Springer Science & Business Media (2013)
5. Bruggi, M.: On an alternative approach to stress constraints relaxation in topology optimization. Structural Multidisciplinary Optimization **36**(2), 125–141 (2008)
6. Chen, Z., Zhou, D., Liao, H., Zhang, X.: Precision alignment of optical fibers based on telecentric stereo microvision. IEEE/ASME Trans. Mechatron. **21**(4), 1924–1934 (2016)
7. Cheng, G., Guo, X.: ε-relaxed approach in structural topology optimization. Structural Optimization **13**(4), 258–266 (1997)
8. De Bona, F., Munteanu, M.G.: Optimized flexural hinges for compliant micromechanisms. Analog Integr. Circ. Sig. Process. **44**(2), 163–174 (2005)
9. De Leon, D.M., Alexandersen, J., Fonseca, J.S., Sigmund, O.: Stress-constrained topology optimization for compliant mechanism design. Structural Multidisciplinary Optimization **52**(5), 929–943 (2015)
10. Holmberg, E., Torstenfelt, B., Klarbring, A.: Stress constrained topology optimization. Structural Multidisciplinary Optimization **48**(1), 33–47 (2013)
11. Hopkins, J.B., Culpepper, M.L.: Synthesis of multi-degree of freedom, parallel flexure system concepts via freedom and constraint topology (fact)-part i: principles. Precis. Eng. **34**(2), 259–270 (2010)
12. Howell, L.L.: Compliant mechanisms. Wiley (2001)
13. Kirsch, U.: On singular topologies in optimum structural design. Structural Optimization **2**(3), 133–142 (1990)
14. Larsen, U.D., Signund, O., Bouwsta, S.: Design and fabrication of compliant micromechanisms and structures with negative poisson's ratio. J. Microelectromech. Syst. **6**(2), 99–106 (1997)
15. Le, C., Norato, J., Bruns, T., Ha, C., Tortorelli, D.: Stress-based topology optimization for continua. Structural Multidisciplinary Optimization **41**(4), 605–620 (2010)
16. Lin, R., Zhang, X., Long, X., Fatikow, S.: Hybrid flexure hinges. Review of scientific instruments **84**(8), 085,004 (2013)
17. Lipkin, H., Patterson, T.: Generalized center of compliance and stiffness. In: Proceedings of 1992 IEEE International Conference on Robotics and Automation, pp. 1251–1256. IEEE (1992)
18. Liu, M., Zhang, X., Fatikow, S.: Topology optimization of large-displacement flexure hinges. In: ASME 2015 International Design Engineering Technical Conferences and Computers and Information in Engineering Conference, pp. V05AT08A005–V05AT08A005. American Society of Mechanical Engineers (2015)
19. Liu, M., Zhang, X., Fatikow, S.: Design and analysis of a high-accuracy flexure hinge. Rev. Scientific Instruments **87**(5), 055,106 (2016)
20. Liu, M., Zhang, X., Fatikow, S.: Design of flexure hinges based on stress-constrained topology optimization. Proc. Inst. Mech. Eng. Part C J. Mech. Eng. Sci. p. 0954406216671346 (2016)
21. Liu, M., Zhang, X., Fatikow, S.: Design and analysis of a multi-notched flexure hinge for compliant mechanisms. Precis. Eng. (2017)

22. Lobontiu, N.: Compliant mechanisms: design of flexure hinges. CRC Press (2002)
23. Lobontiu, N., Paine, J.S., Garcia, E., Goldfarb, M.: Corner-filleted flexure hinges. J. Mech. Des. **123**(3), 346–352 (2001)
24. Lobontiu, N., Paine, J.S., Garcia, E., Goldfarb, M.: Design of symmetric conic-section flexure hinges based on closed-form compliance equations. Mech. Mach. Theory **37**(5), 477–498 (2002)
25. Lobontiu, N., Paine, J.S., OMalley, E., Samuelson, M.: Parabolic and hyperbolic flexure hinges: flexibility, motion precision and stress characterization based on compliance closed-form equations. Precis. Eng. **26**(2), 183–192 (2002)
26. Paros, J.M.: How to design flexure hinges. Mach. Des. **37**, 151–156 (1965)
27. Pilkey, W.D., Pilkey, D.F.: Peterson's Stress Concentration Factors. Wiley (2008)
28. Rozvany, G., Sobieszczanski-Sobieski, J.: New optimality criteria methods: forcing uniqueness of the adjoint strains by corner-rounding at constraint intersections. Structural Optimization **4**(3–4), 244–246 (1992)
29. Schotborgh, W.O., Kokkeler, F.G., Tragter, H., van Houten, F.J.: Dimensionless design graphs for flexure elements and a comparison between three flexure elements. Precis. Eng. **29**(1), 41–47 (2005)
30. Sigmund, O.: On the design of compliant mechanisms using topology optimization. Mech. Structures Mach. **25**(4), 493–524 (1997)
31. Smith, S.T.: Flexures: Elements of Elastic Mechanisms. CRC Press (2000)
32. Smith, S.T., Badami, V.G., Dale, J.S., Xu, Y.: Elliptical flexure hinges. Rev. Sci. Instrum. **68**(3), 1474–1483 (1997)
33. Su, H.J., Dorozhkin, D.V., Vance, J.M.: A screw theory approach for the conceptual design of flexible joints for compliant mechanisms. J. Mech. Robot. **1**(4), 041,009 (2009)
34. Suzuki, K., Kikuchi, N.: A homogenization method for shape and topology optimization. Comput. Methods Appl. Mech. Eng. **93**(3), 291–318 (1991)
35. Svanberg, K.: The method of moving asymptotesła new method for structural optimization. Int. J. Numer. Meth. Eng. **24**(2), 359–373 (1987)
36. Tian, Y., Shirinzadeh, B., Zhang, D.: Closed-form compliance equations of filleted v-shaped flexure hinges for compliant mechanism design. Precis. Eng. **34**(3), 408–418 (2010)
37. Trease, B.P., Moon, Y.M., Kota, S.: Design of large-displacement compliant joints. J. Mech. Des. **127**(4), 788–798 (2005)
38. Tseytlin, Y.M.: Notch flexure hinges: an effective theory. Rev. Sci. Instrum. **73**(9), 3363–3368 (2002)
39. Wang, M.Y., Wang, X., Guo, D.: A level set method for structural topology optimization. Comput. Methods Appl. Mech. Eng. **192**(1), 227–246 (2003)
40. Wang, R., Zhang, X.: A planar 3-DOF nanopositioning platform with large magnification. Precis. Eng. **46**, 221–231 (2016)
41. Xie, Y.M., Steven, G.P.: A simple evolutionary procedure for structural optimization. Comput. Structures **49**(5), 885–896 (1993)
42. Xu, W., King, T.: Flexure hinges for piezoactuator displacement amplifiers: flexibility, accuracy, and stress considerations. Precis. Eng. **19**(1), 4–10 (1996)
43. Yong, Y.K., Lu, T.F., Handley, D.C.: Review of circular flexure hinge design equations and derivation of empirical formulations. Precis. Eng. **32**(2), 63–70 (2008)
44. Yu, J., Li, S., Su, H.j., Culpepper, M.: Screw theory based methodology for the deterministic type synthesis of flexure mechanisms. J. Mech. Robot. **3**(3), 031,008 (2011)
45. Zelenika, S., Munteanu, M.G., De Bona, F.: Optimized flexural hinge shapes for microsystems and high-precision applications. Mech. Mach. Theory **44**(10), 1826–1839 (2009)
46. Zhu, B., Zhang, X., Fatikow, S.: Design of single-axis flexure hinges using continuum topology optimization method. Sci. China Technol. Sci. **57**(3), 560–567 (2014)

Chapter 3
Topology Optimization of Distributed Compliant Mechanisms

Abstract This chapter focuses on the point flexure problem when using continuum topology optimization methods to design compliant mechanisms. Developing alternative optimization models is adopted to solve the de facto hinge problem aimed at obtaining compliant mechanisms with distributed compliance. Two methods are developed in this chapter and several numerical examples are presented to demonstrated the validity of the proposed methods. Extensions of the proposed methods for designing compliant mechanisms with multiple outputs are also discussed.

3.1 Introduction

For designing compliant mechanisms with single input-output behavior, the design domain can be illustrated in Fig. 3.1 where Γ_d indicates the Dirichlet boundary. An input force F_{in} is applied at the input port i. Δ_{in} and Δ_{out} indicate the displacements occurred at the input port i and the output port o due to F_{in}, respectively. A spring with stiffness k_{out} is attached to the output port to imitate the reaction force from the workpiece by $F_{out} = k_{out}\Delta_{out}$ [2].

Similar to many engineering design problems, the design of compliant mechanisms can be posed as an optimization problem with the following statement:

$$\begin{aligned} \min: &\quad J(x) \\ \text{s.t.}: &\quad g(x) \leq 0 \end{aligned} \tag{3.1}$$

where x is the design variable which can describe the competing design candidates; J is the objective function; the inequality $g(x) \leq 0$ is called constraint which imposes specified requirements on the design variable x.

It is well-known that, for topology optimization of the stiffest structures, there is a universally accepted formulation (minimizing the mean compliance subject to a constraint on the volume of material used [2]). However, a universally accepted formulation does not exist for the topology optimization of compliant mechanisms, although several available formulations have been developed over the past decades [5].

X. Zhang and B. Zhu, *Topology Optimization of Compliant Mechanisms*,
https://doi.org/10.1007/978-981-13-0432-3_3

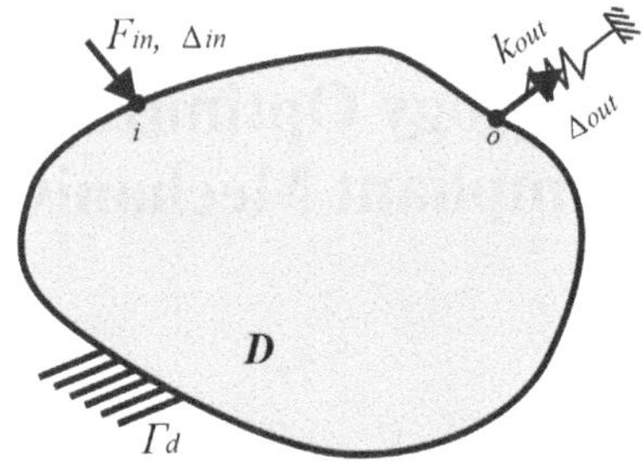

Fig. 3.1 The design domain of the compliant mechanisms with single input single output behaviour

A rough collection of the established objective functions for designing compliant mechanisms is shown in Table 3.1 in which *MSE* is the mutual strain energy which dedicates the flexibility of the designed mechanism; *SE* is the strain energy which dedicates the stiffness; ω is the weighting factor; *MA* is the mechanical advantage; *GA* is the geometrical advantage, and *ME* is the mechanical efficiency.

These established formulations can be studied under two main groups, the first of which is established by maximizing some kind of mechanical measurements. For example, Sigmund [24] proposed a formulation in which the mechanical advantage is employed as the objective function. In his method, the maximum input displacement is also limited for indirectly controlling the maximum stress level. Some available formulations based on other mechanical measurements, such as geometrical advantage [14], mechanical efficiency [13] and output displacement [2] have also been developed.

The second of which groups is formulated by treating the design as a multi-objective problem where flexibility and compliance are both considered to meet the function requirement and the strength requirement of a compliant mechanism. For example, Ananthasuresh [1] proposed a weighted linear combination formulation to accomplish simultaneously the two requirements of both flexibility (maximizing mutual potential energy) and compliance (minimizing strain energy). However, choosing the value of the normalized weight in the global sense is rather difficult. To overcome this limitation, Frecker et al. [7] proposed an alternative multi-criteria objective of maximizing the ratio of the flexibility requirement and the mean compliance. Nishiwaki et al. [18] considered multiple objectives in the optimization of compliant mechanisms by using the homogenization method, but found that the use of the homogenization method is somewhat troublesome. Saxena and Ananthasuresh [23] proposed an energy based method to accomplish two requirements of the compliant mechanisms topology optimization by maximizing the flexibility, minimizing the compliance and maximizing the mechanical advantage of the mechanism simultaneously. In addition, Luo et al. [17] proposed a new multi-objective method for the topology optimization of compliant mechanisms, in which a hybrid-filtering scheme is developed for solving numerical instabilities. It was shown that the scheme can also prevent one node connected hinges in the design of compliant mechanisms. A comparative study of different objective functions for topology optimization of compliant mechanisms can be found in [5].

Table 3.1 Established objective functions for designing compliant mechanisms

No.	Authors	Objective function J
1	Ananthasuresh [1]	$-\omega MSE + (1-\omega)SE$
2	Frecker et al. [7]	$-\frac{MSE}{SE}$
3	Nishiwaki et al. [18], Joo et al. [11]	$-\frac{MSE}{\omega SE_i+(1-\omega)SE_o}$
4	Kota et al. [13]	$-\frac{GA}{SE}$
5	Saxena and Ananthasuresh [23]	$-sign(MSE)\frac{\frac{1}{2}k_{out}MSE^2}{SE}$
6	Bendsøe and Sigmund [2]	$-\Delta_{out}$
7	Sigmund [24]	$-MA$
8	Lau et al. [14], Hetrick [9]	$-ME$
9	Chen [3], Lau et al. [14]	$-GA$

3.2 Point Flexures

In the context of designing compliant mechanisms using continuum topology optimization methods, one difficulty is the tendency for de facto hinges (or point flexures, see Fig. 3.2) to appear in the obtained compliant mechanisms [21].

Compliant mechanisms containing de facto hinges are not truly compliant but rather they evolve to rigid-link mechanisms. For this reason, compliant mechanisms containing de facto hinges belong to the lumped compliant mechanisms since the flexibility of the mechanisms is provided only in the de facto hinge regions.

The causes of de facto hinges have been studied from the viewpoint of maximizing energy transfer, as in [3, 16]. Indeed, most of the formulations for designing compliant mechanisms are force-deflection type and they attempt to simultaneously maximize the deformation at the output port and maximize the overall stiffness of the structure which is measured by using the strain energy. As a result, a rigid-body linkage with revolute joints is the true optimum since it can generate large output deformation and has a zero strain energy. Thus, the de facto hinges are inevitable.

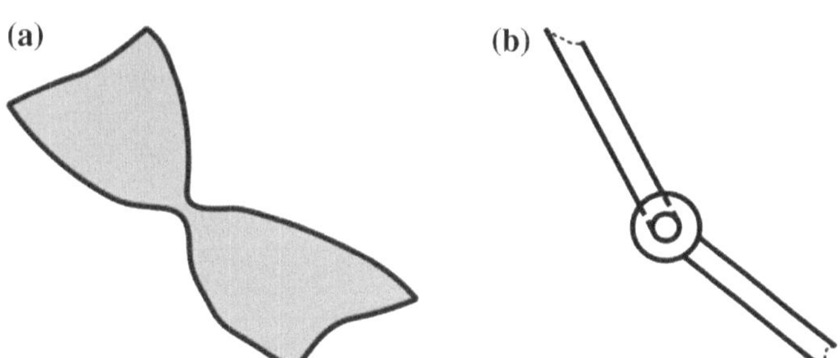

Fig. 3.2 **a** de facto hinge in compliant mechanisms, and **b** revolute joint of the rigid body

One of the biggest shortcomings is that the stress in the sharp hinge regions could approach infinity and the structure would break in the area of the de facto hinges. Moreover, it is very difficult to fabricate such de facto hinges for micro-scale mechanical systems. Therefore, for most real-world applications, hinge-free compliant mechanisms are preferable.

Great efforts have been made by various researchers who attempted to remove the de facto hinges to obtain compliant mechanisms with distributed compliance. Extra constrains (e.g., the slope constrained method [19], the stress constraints method [6], the minimum length scale constraint method [20]) and the filtering methods (e.g., the hybrid-filtering method [17] and the morphology-based filtering method [26]) are often employed into the optimization process. Suppressing the one-node connected hinges by additional constraints has shown some success. However, as pointed in [22], it may result in considerable weight increases. Moreover, all these schemes are fundamentally second stage operations or local redesign procedures and appear to have a limit success on preventing de facto hinges [27].

Another notable approach is to use the second stage design operation strategy to redesign the de facto hinge regions as continuous material bridges. The main shortcoming of this method is that the obtained mechanisms may deviate from the original design. Hybrid Discretization method [12] and polygonal mesh method [31] have been developed for eliminating the point connections as well as the checkerboard problem. These methods require re-meshing or use high order element of the design domain and therefore they are computationally expensive.

The reasons for point flexures in monolithic compliant mechanisms topology optimization might be still in debate up-to-date. The most compelling reason of causing de facto hinges lied behind the mathematical formulations [27]. Therefore, developing alternative formulations for compliant mechanisms may be the most efficient way to avoid de facto hinges. To this end, many formulations have been proposed aimed at essentially preventing de facto hinges and some of them are listed in Table 3.2 where Ξ is the quadratic energy functional; $GA^\star$ is the target geometrical advantage; $\mathbf{K}_{11}$ and $\mathbf{K}_{22}$ are the characteristic stiffness; $\bar{\varepsilon}$ is the effective strain, and F_1 and F_2 are two class functions in the physical programming method.

One may notices that wang and Chen [28] proposed an intrinsic characteristic stiffness (CS) method based on the level set method. Reference [5] pointed out that the CS method gives compliant mechanisms with de facto hinges for a large desired output performance. In addition, via two different sets of artificial spring models,

Table 3.2 Established objective functions for avoiding de facto hinges

No.	Authors	Objective function J
1	Luo et al. [16]	$-ME + \omega\Xi$
2	Wang and Chen [28]	$-e^{(GA-GA^*)^2}\mathbf{K}_{11}\mathbf{K}_{22}$
3	Lee and Gea [15]	$F_1(\sum_{i=1}^{n}\bar{\varepsilon}_i^2) + F_2 \cdot MSE$

Rahmatalla and Swan [21] proposed a new method for designing hinge-free compliant mechanisms. Incorporating with two types of mean compliances, Zhu and Zhang [32] developed several objective formulations for topology optimization distributed compliant mechanisms incorporating with the level set method. The method can be used for preventing de facto hinges so that distributed compliant mechanisms can be obtained. The method has been further extended for topology optimization of compliant mechanisms with multi-outputs [33].

In the following sections, we are motivated to develop alternative mathematical formulations of the topology optimization of compliant mechanisms aiming at automatically implying the absence of the de facto hinges.

3.3 A Method of Reducing Two Output Performances

In order to illustrate the underlying idea of the method, suppose that a compliant mechanism with de facto hinges is shown in Fig. 3.3. We consider two loading conditions which are shown in Fig. 3.3a and b, respectively. The difference is that in the second condition, k_{out} is removed.

Suppose that the output displacement Δ_{out} is regarded as the output performance p. When the mechanism is loaded with a force F_{in}, a displacement Δ_{o1} can be obtained at the output port from the first condition. Similarly, a displacement Δ_{o2} can be obtained from the second condition due to the same load F_{in}.

Let us define a parameter u_Δ as follows

$$u_\Delta = |\Delta_{o1} - \Delta_{o2}|. \tag{3.2}$$

For a hinged compliant mechanism, u_Δ will be extremely large because of the large Δ_{o2} (without being constrained by the spring, the hinge part undergoes an essential rigid-body rotation).

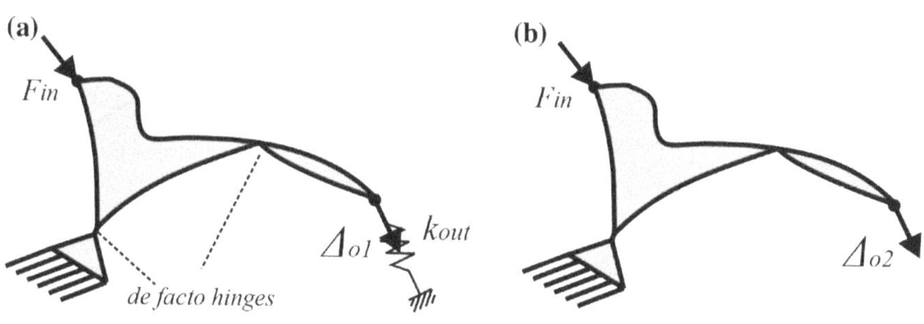

Fig. 3.3 A hypothetic hinged-compliant mechanism: **a** with a spring attached to its output port and **b** with the spring removed

If the de facto hinges are replaced with distributed hinges, u_Δ will become smaller. The reason is that the distributed hinges are equivalent to torsion springs [10]. For the loading condition in Fig. 3.3b, Δ_{o2} will become smaller because the distributed hinges will store a certain level of energy. This means that de facto hinges correspond to a large u_Δ, whereas distributed hinges correspond to a small one. Conversely, by constraining this type of difference, de facto hinges should be prevented.

If we treat the spring that is attached to the output port as a barrier, reducing the difference is actually equivalent to lowering the effectiveness of the barrier. If the output performance of a mechanism is hardly changed whether a barrier exists at its output port, the mechanism has to be stiff enough (free of de facto hinges, which make the mechanism very fragile).

For simplicity, we use the square of the difference instead of the absolute value to represent the performance difference. A more general modeling method for the topology optimization of hinge-free compliant mechanisms can be stated as follows: in addition to maximizing the performance p, p_Δ needs to be minimized during the optimization process, where p_Δ can be stated as

$$p_\Delta = \frac{1}{2}(p_1 - p_2)^2 \tag{3.3}$$

where p_1 is the performance that is obtained from the first elastic analysis where a spring is attached to the output port, and p_2 is the performance that is obtained from the second elastic analysis where the spring is eliminated (Fig. 3.4).

Therefore, a multi-objective optimization problem is presented. Many methods can be used to transform the multi-objective problem into a single objective optimization problem to simplify the optimization process, such as the weighted sum method and the ratio method. We adopt the weighted sum method due to its ease of use. Therefore, an alternative objective function can be stated as

$$\min: \quad J = -(1-\omega)p + \omega p_\Delta \tag{3.4}$$

where $0 \leq \omega \leq 1$ is the weighting factor.

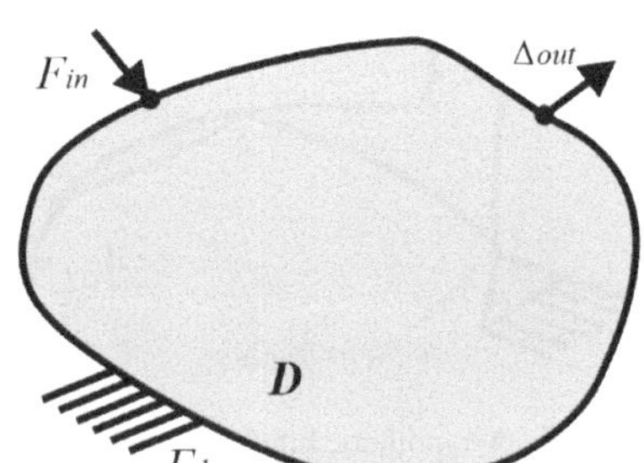

Fig. 3.4 The boundary conditions for the second elastic analysis

3.3.1 A GA-Based Formulation

One can simply set GA to illustrate the compliant mechanism performance measure:

$$p = GA = \frac{\Delta_{out}}{\Delta_{in}} \tag{3.5}$$

Using the dummy load method [24], Δ_{in} and Δ_{out} can be further expressed as follows:

$$\Delta_{in} = \frac{-F_{in}u_{1i} - F_{in}k_{out}u_{1o}u_{2i} + F_{in}k_{out}u_{1i}u_{2o}}{-1 + k_{out}u_{2o}} \tag{3.6}$$

$$\Delta_{out} = \frac{-F_{in}u_{1o}}{-1 + k_{out}u_{2o}} \tag{3.7}$$

where u_{1i} and u_{1o} are the displacements at the input and output port, respectively, which are included in the displacement vector u_1. u_{2i} and u_{2o} are the displacements at the input and output port, respectively, which are included in the displacement vector u_2. Note that u_1 and u_2 are obtained by solving two self-adjoint weak forms corresponding to the two unit-loading cases f_1 and f_2, respectively.

One may use GA_2 to denote the geometrical advantage obtained from the second elastic analysis. GA_2 can be defined as

$$GA_2 = \frac{u_{1o}}{u_{1i}} \tag{3.8}$$

and the difference between the two geometrical advantages is stated as follows:

$$GA_{\Delta} = \frac{1}{2}(GA - GA_2)^2 \tag{3.9}$$

3.3.1.1 Optimization Model

A possible optimization formulation can be expressed as follows. Note that a minus is used because the maximization of GA is equivalent to the minimization of $-GA$.

$$\begin{aligned} Q_N: \quad & \min: \; J = -(1-\omega)GA + \omega GA_{\Delta} \\ & \text{s.t.}: \; Vol \le Vol^{max} \\ & \quad\quad \Delta_{in} \le \Delta_{in}^{max} \\ & \quad\quad a(u, v) = l(v), \quad \forall v \in U \end{aligned} \tag{3.10}$$

where Vol is the maximum allowable material usage which is limited by an upper value Vol^{max}, Δ_{in} is the input displacement due to the input load and is limited by Δ_{in}^{max}. The state equation is expressed in terms of the energy bilinear functional $a(u, v)$ and the load linear functional $l(v)$ as

$$\int_D E_{ijkl}\varepsilon_{ij}(u)\varepsilon_{kl}(v)d\Omega = \int_D bvd\Omega + \int_\Gamma \tau vdS \tag{3.11}$$

where E_{ijkl} denotes the stiffness tensor. b and τ denote the body force and boundary traction, respectively. U denotes the space of the kinematically admissible displacement fields, v denotes the arbitrary virtual displacement in the space U. If ω is set to 0, Q_N will reduce to

$$\begin{aligned} Q_T: \min: \quad & J = -GA \\ \text{s.t.}: \quad & Vol \leq Vol^{max} \\ & \Delta_{in} \leq \Delta_{in}^{max} \\ & a(u, v) = l(v), \qquad \forall v \in U \end{aligned} \tag{3.12}$$

which is regarded as a traditional model for the topology optimization of compliant mechanisms [3].

For implementation of Q_N, although the GA has to be calculated twice in each optimization iteration step, the finite element analysis (FEA), which is regarded as the most time-consuming step in each iteration [32], only needs to be performed once (this can be seen directly from Eqs. (3.5) and (3.8)). Therefore, adding the second performance calculation does not occupy much CPU time.

3.3.1.2 Validity

In order to examine the validity of the optimization formulation Q_N on preventing the de facto hinges, the displacement inverter is considered. This example has been widely studied and regarded as a benchmark problem in the literature of topology optimization of compliant mechanisms. The level set method is employed for solving Q_N to obtain the optimal solutions.

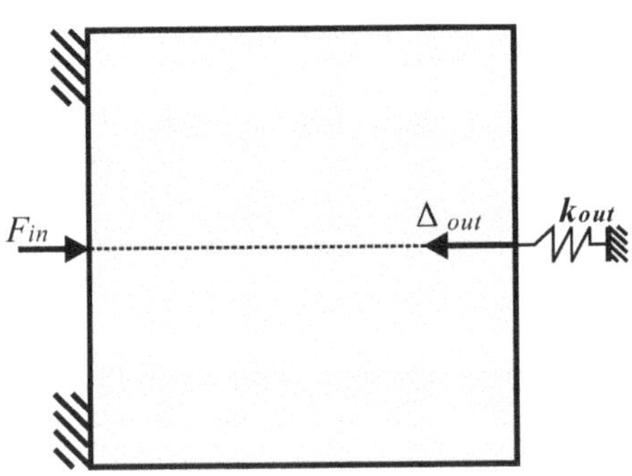

Fig. 3.5 The design domain of the displacement inverter

The design domain is shown in Fig. 3.5 and the goal of the design is to get a mechanism that can convert the input displacement on the left edge of the design domain to the displacement in the opposite direction on the right edge. Due to symmetry, only half of the design domain is taken into consideration for elastic analysis. However, in order to show a clear result, the final topologies are all shown entirely.

Effect of ω In order to examine the effect of the ω on the optimal topology, six cases are studied in which the ω is set to 0, 0.2, 0.4, 0.6, 0.8 and 0.9.

The obtained final designs are shown in Fig. 3.6. The computed performance characteristics of such inverter designs are shown in Table 3.3 where N denotes the total iterations. The de facto hinge problem is inevitable when Q_T is used.

From the schematic drawing its deformation, one can see that the material regions (structural members) around the hinge parts undergo essential rigid-body motion. The final design that is shown in Fig. 3.6a confirms that the de facto hinge problem is inevitable when the traditional optimization model is used.

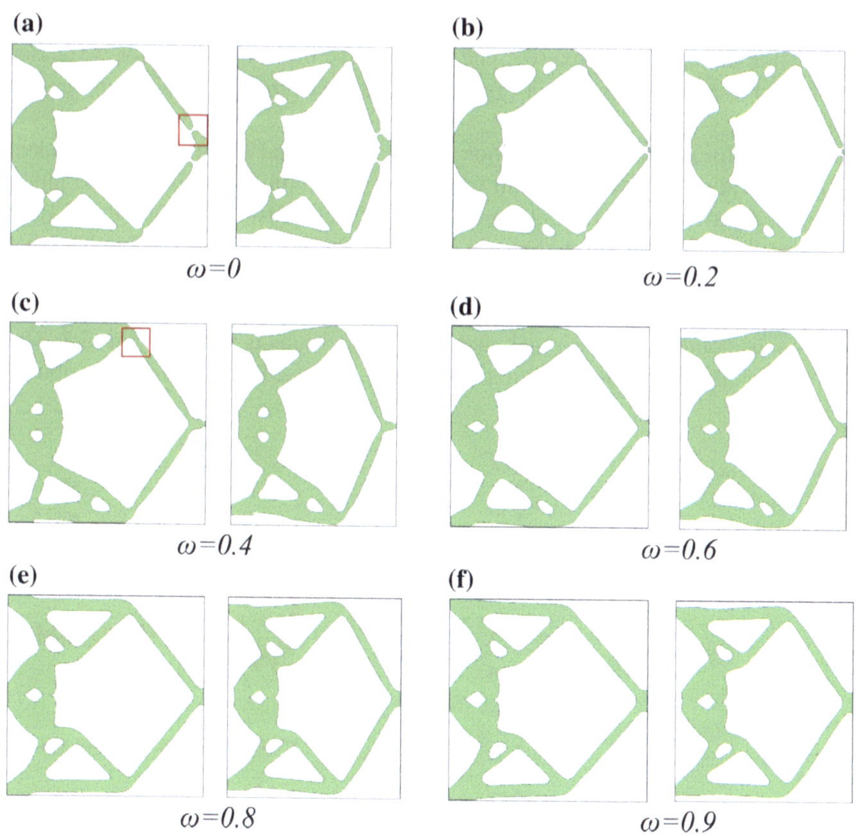

Fig. 3.6 The final designs of the displacement inverter obtained using the optimization model Q_N with different ω

Table 3.3 The computed performance characteristics of the displacement inverter designs at finite deformation

ω	N	GA	GA_2	GA_Δ
0	264	1.1251	1.6317	0.1283
0.2	252	1.1145	1.5617	0.1000
0.4	304	1.0513	1.4396	0.0754
0.6	249	0.9533	1.2509	0.0443
0.8	345	0.7800	0.9632	0.0168
0.9	260	0.7106	0.8578	0.0108

In cases of ω is set to be quite large, e.g., 0.4, 0.6, 0.8 and 0.9, the obtained devices are all free of de facto hinges. From the computed performance characteristics of such devices, one can confirm that by reducing the performance difference GA_Δ, the de facto hinges can be successfully prevented. The respective deformed configurations of these devices show the mechanisms functioning with well-distributed elastic deformation and no point flexures. The convergence history of GA_Δ for two cases ($\omega = 0$ and $\omega = 0.6$) is shown in Fig. 3.7 in which a decrease of GA_Δ can be seen directly.

However, for using the optimization model Q_N, special attention must be taken into consideration since with a small value of the weighting factor ω, it could fail to prevent de facto hinges in the obtained compliant mechanism. Although the GA_Δ is also decreased compared with the case of $\omega = 0$, it is not decreased enough so that a hinge-free inverter can be obtained.

In Fig. 3.6, note that the hinge regions are marked with the red squares in particular as an example. A close look at these hinge regions is shown in Fig. 3.8. The traditional method gives de facto hinges where the proposed method gives distributed

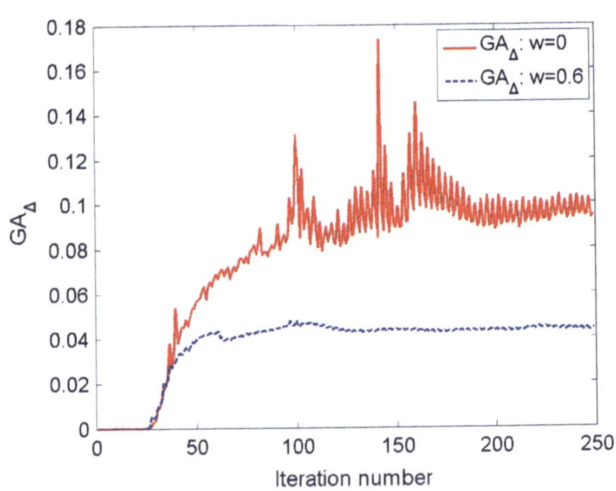

Fig. 3.7 The convergence history of GA_Δ of the displacement inverter using Q_T ($\omega = 0$) and Q_N ($\omega = 0.6$)

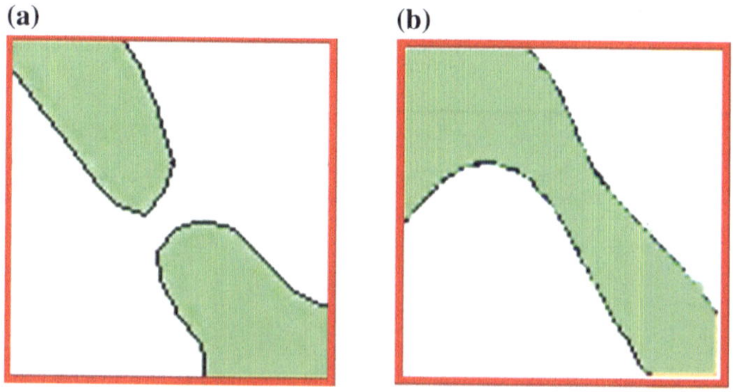

Fig. 3.8 A close look at the hinges in Fig. 3.6. The traditional approach Q_T produces de facto hinges where the two-step elastic modeling method Q_N with $\omega = 0.4$ gives distributed hinges: **a** hinge region in Fig. 3.6a; **b** hinge region in Fig. 3.6c

hinges. Note also that the position of the hinges is also different which highlights that the results obtained by using the present method could not be obtained by simply employing the second design stage method.

Achieving Mechanisms with High Performance The performance characteristics of the hinged device appear to be remarkably better than those of hinge-free device, see Table 3.3. The elimination of de facto hinges can also decrease the output performances. However, it must be borne in mind that the designs that contain de facto hinges are too fragile to be loaded. Thus the performance characteristics of the compliant mechanisms that rely on de facto hinges are not necessarily achievable.

To obtain a hinge-free compliant mechanism that has higher output performances, one can impose more restrictive material usage constraint values. In order to demonstrate this idea, a redirector mechanism is further considered. The design domain is shown in Fig. 3.9a. Part of its left side is fixed and an external force is applied in the middle of its left side. Two vertical output displacements are expected at the upper and lower corners of its right side. The design domain is defined within a 100×100 square.

When the maximal material usage is restricted to 0.3, Q_T gives a hinged result as shown in Fig. 3.9b with the geometrical advantage $GA = 2.65$. Q_N gives a hinge-free result as shown in Fig. 3.9c with a smaller geometrical advantage $GA = 2.49$.

If one use Q_N and a smaller maximal material usage, such as 0.2, a hinge-free displacement redirector with $GA = 2.64$ can be obtained. The final topology, as well as its deformed configuration, is shown in Fig. 3.9d. The obtained mechanism appears to have comparable output performance with the device shown in Fig. 3.9b while being completely free of de facto hinges.

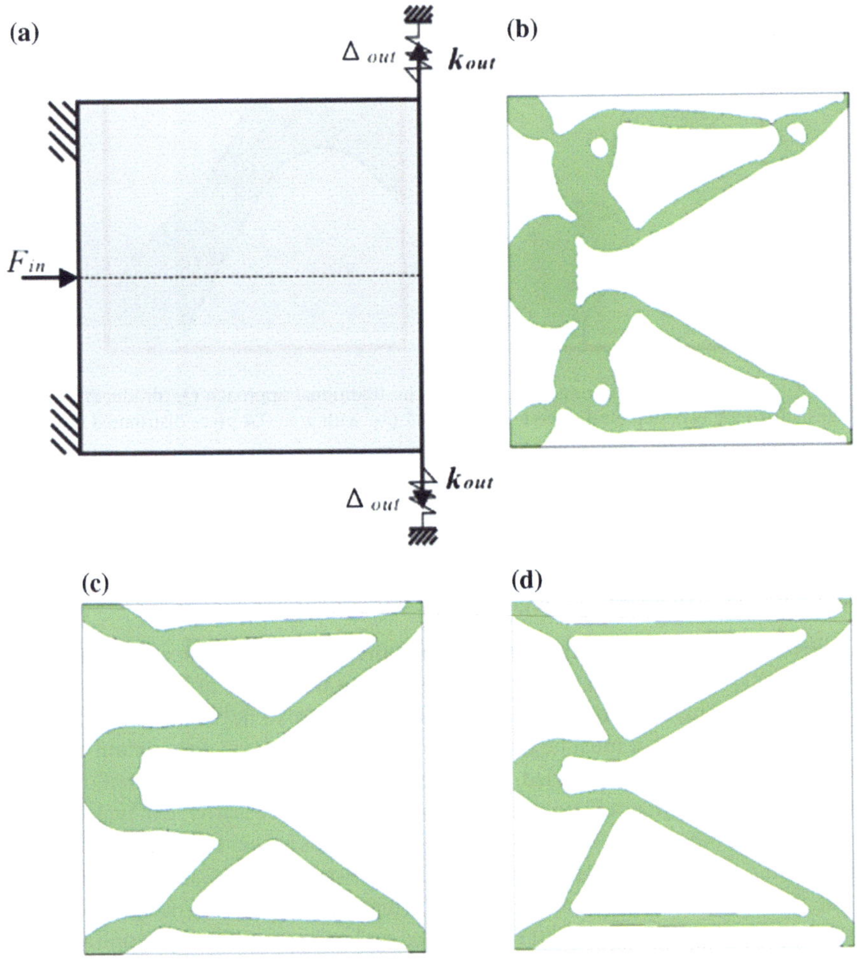

Fig. 3.9 **a** The design domain of the displacement redirector; The final topologies of the displacement redirector obtained using **b** Q_T, and **c** Q_N; **d** The final design of the displacement redirector with $Vol^{max} = 0.2$

3.3.2 A Self-adjust Scheme for Setting Weighting Factors

By setting an appropriate ω, Eq. (3.4) can be available for all kinds of design cases even when a small k_{out} is used. The major problem of the Eq. (3.4) is that the ω can not be chosen in the global sense. The ω may be varying from one problem to anther. One may need to try several times to find the appropriate value.

To avoid specifying the weighting factor, one possible way is to formulate the optimization problem by using the ratio method, i.e., the objective that should be minimized can be set as follows:

$$J = \frac{f(p_\Delta)}{g(p)} \tag{3.13}$$

where f and g are the increasing functions of p_Δ and p, respectively. However, the appropriate setting of f and g is also uneasy.

For the implementation of the proposed method, e.g., Eq. (3.4), to avoid specifying the weighting factor, a self-adjust weighting factor should be more suitable. The weighting factor can be set according to the principles as follows: It can dynamically change upon the optimization process meanwhile preventing p_Δ and p dominating each other.

To this end, it is useful to use the information in the previous iteration. The objective function Eq. (3.4) can be rewritten as

$$\min: \quad J = -(1-\omega^k)p + \omega^k p_\Delta \tag{3.14}$$

where ω^k denotes the value of weighting factor ω at the kth iteration. It changes with each iteration k of the optimization algorithm with

$$\frac{\omega^k}{1-\omega^k} = |\frac{p^{k-1}}{p_\Delta^{k-1}}|, \quad k = 1, 2, \ldots n \tag{3.15}$$

where p^{k-1} and p_Δ^{k-1} are the values of p and p_Δ at iteration $k-1$. Equation (3.15) can be further written as

$$\omega^k = \frac{1}{1 + \left|\frac{p_\Delta^{k-1}}{p^{k-1}}\right|}, \quad k = 1, 2, \ldots n \tag{3.16}$$

The reason of setting ω based on the information at the previous step is that it can avoid p and p_Δ going insanely large. Based on the weighting factor setting scheme, if p becomes larger, then ω will become larger. The weight of p_Δ is increased and thus they make the created mechanism stiffer. Conversely, if p becomes smaller, then ω will become smaller. This will decrease the weight of p_Δ to make the created mechanism more flexible.

From the above analysis one can see that there is no need for manually operating the functions because the weighting factors can actually self-adjust based on the information obtained from the previous iteration. If an optimization problem has n optimization steps before convergence, it has n objective functions to deal with. However, during each iteration, the ω is actually a constant. Therefore

$$\frac{\partial \omega}{\partial x} = 0 \tag{3.17}$$

where x is the design variable. This makes the sensitivity of the proposed method easy to compute.

3.3.2.1 Optimization Model

In order to demonstrate the validity of Eq. (3.14), one can simply set p to the output displacement Δ_{out}. Using the proposed reducing output performances methods, an alternative optimization model Q_{Nsa} can be formulated as follows.

$$\begin{aligned} Q_{Nsa}: \quad & \min: \quad J = \omega u_{\Delta} - (1-\omega)\Delta_{o1} \\ & \text{s.t.}: \quad Vol \leq Vol^{max} \\ & \qquad\quad \Delta_{in} \leq \Delta_{in}^{max} \end{aligned} \tag{3.18}$$

where Δ_{o1} is the output displacement that is obtained using the first boundary condition (see Fig. 3.1), u_{Δ} is defined using Eq. (3.2). ω is the weighting factor which changes along with the optimization process

$$\omega^k = \frac{1}{1 + \left| \frac{u^{k-1}\Delta}{\Delta_{o1}^{k-1}} \right|}, \quad k = 1, 2, \ldots n \tag{3.19}$$

With a decreasing of u_{Δ} and an increasing of Δ_{o1}, ω will be gradually enlarged. It means the weight of u_{Δ} is increased. This can minimize u_{Δ} to prevent de facto hinges meanwhile avoiding setting ω artificially.

Incorporating with the SIMP method, the optimization problem can be rewritten by using the finite element method form as follows

$$\min_{\mathbf{x}}: \quad J = \omega|\mathbf{L}^T\mathbf{U}_{out}^a - \mathbf{L}^T\mathbf{U}_{out}^b| - (1-\omega)\mathbf{L}^T\mathbf{U}_{out}^b \tag{3.20}$$

$$\text{s.t.}: \quad \sum_{e=1}^{N} v_e x_e \leq Vol^{\max} \tag{3.21}$$

$$\Delta_{in} \leq \Delta_{in}^{max} \tag{3.22}$$

$$\mathbf{K}_a\mathbf{U}_{out}^a = \mathbf{F}_{in} \tag{3.23}$$

$$\mathbf{K}_b\mathbf{U}_{out}^b = \mathbf{F}_{in} \tag{3.24}$$

$$0 < x_{\min} \leq x_e \leq 1 \tag{3.25}$$

$$e = 1, 2, \ldots N \tag{3.26}$$

where $\mathbf{x}$ is the design variable matrix. $\mathbf{L}$ is a unit load vector consisting of zero except for position i with an entry of $\mathbf{L}_i = 1$, and $\mathbf{U}_{out}^a$ is the displacement vector generated by the load vector $\mathbf{F}_{in}$ applied at the input port with a spring fixed at the output port of the design domain. $\mathbf{U}_{out}^b$ is the displacement vector generated by the load vector $\mathbf{F}_{in}$ applied at the input port without a spring fixed at the output port of the design domain. The stiffness matrixes $\mathbf{K}_a$ and $\mathbf{K}_b$ depend on the stiffness E_e in element e which can be written in the form

$$\mathbf{K}_a = \sum_e^N \mathbf{K}_e(E_e) = \mathbf{K} \tag{3.27}$$

$$\mathbf{K}_b = \sum_e^N \mathbf{K}_e(E_e) + \mathbf{K}_{out} = \mathbf{K} + \mathbf{K}_{out} \tag{3.28}$$

where $\mathbf{K}_e$ is the element stiffness matrix in the global level. $\mathbf{K}_{out}$ is the artificial spring stiffness k_{out} in the global level.

3.3.2.2 Optimization Algorithm

During each optimization iteration, the finite element analysis step is the most CPU time consuming step. Further, during each finite element analysis step, most of the computational efforts are invested in the establishment of the stiffness matrix $\mathbf{K}$. In the implementation of the proposed method, although an extra equilibrium equation Eq. (3.23) needs to be solved, the stiffness matrix $\mathbf{K}$ is only needed to be invested once. Therefore, compared with the traditional formulation, the proposed PDR method guarantees hinge-free topologies without losing the computational efficiency.

The sensitivity analysis can be easily accomplished using the adjoint method [2] and details are not given here because they can be easily found in the literature. For implementation of the SIMP method, numerical instabilities, e.g., checkerboards, will occur during the optimization. Additional regularization methods, such as the perimeter control method [8] and the filtering method [2], are needed to be added to augment the original formulation. The filtering method used in [25], which can directly limit the variations of the densities and/or sensitivities by smoothing the gradients, is employed to prevent the checkerboard patterns.

The SIMP approach leads to a sizing problem and the number of design variables is typically very large. This optimization problem can be efficiently solved by gradient-based optimization algorithms. Since the optimization problem has just few constraints in addition to a simple constraint giving upper and lower limits on the density variable, optimal criteria (OC) methods that have been successfully applied in [25] are employed to solve the optimization problem.

Therefore, the flowchart of the optimization procedure can be shown in Fig. 3.10. In the first step, design parameters, such as the design variable $\mathbf{x}$ that represents the intimal configuration, are given. In the second step, the stiffness matrix $\mathbf{K}$ is built. In the third step, the first equilibrium equation is solved. In the fourth step, the second equilibrium equation is solved. In the fifth step, the sensitivity analysis is performed. The sensitivities are filtered to prevent the checkerboard patterns. The design variables are updated by using the OC method in the sixth step. After a

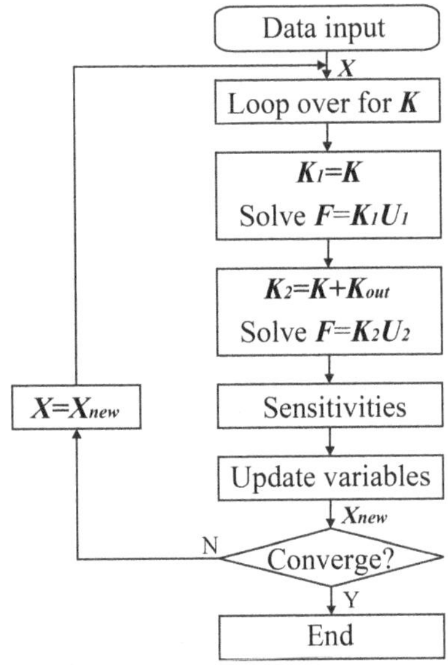

Fig. 3.10 The flow chart of using Q_{Nsa} to topological design compliant mechanisms

convergence check, optimization process is repeated from the second step to the seven step.

3.3.2.3 Numerical Example

The set design computation results presented in this section are intended to highlight the effect of the optimization model Q_{Nsa} on preventing de facto hinges and obtaining compliant mechanisms with distributed compliance. To this end, the displacement inverter problem is resolved.

The maximal material usage is restricted to 0.2 of the whole material that the design domain can occupy. k_{out} is set to be $10^{-3}k_b$ where k_b is the *bounding spring value* and is computed by applying a unit load to the mechanism at the input port with the structural domain fully occupied by the structural material, see [21].

Comparison of the Final Topologies The optimized configurations and their deformed configurations obtained by using the traditional formulation [4] and Q_{Nsa} with the same spring k_{out} are shown in Fig. 3.11. The traditional method gives lumped compliant mechanism while the proposed method gives distributed compliant mechanism. The capability of Q_{Nsa} for hinges suppression is feasible.

Note that the obtained topologies are obviously different from each other which highlights that the results obtained by using the present method could not be obtained by simply employing the second design stage method. This can be important for the reliability and manufacture of the obtained compliant mechanisms. The deformed configuration shown in Fig. 3.11a reveals that when the mechanism with de facto

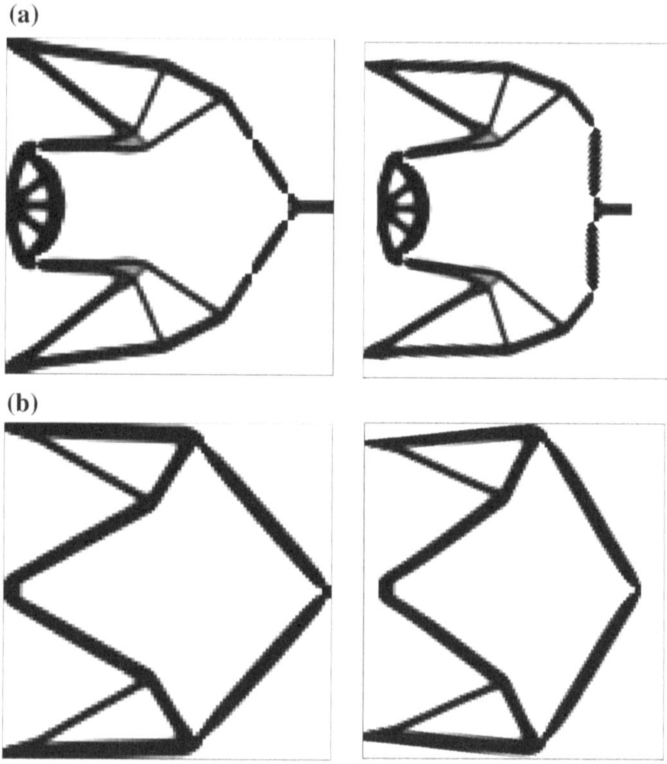

Fig. 3.11 The optimized topologies of the displacement inverter are obtained by using: **a** the traditional formulation, and **b** Q_{Nsa}. The same volume ratio, input load and output spring are used

hinges is loaded, the surrounding materials (structural members) undergo essentially rigid-body rotations. Since it is difficult to fabricate reliable hinges in micro-scale, designs with de facto hinges are of questionable utility. The deformed configuration shown in Fig. 3.11b shows that the hinge-free mechanism functioning with well-distributed elastic deformation which is suitable for real world applications.

For implementation of Q_{Nsa}, ω is dynamically changed during the optimization process. The convergence history of ω is shown in Fig. 3.12. It reveals that, with the increasing of Δ_{out} and the decreasing of Δ_u, ω can vary dynamically and be enlarged and thus the lumped compliance is prevented.

Computational Efficiency In order to verify the efficiency of the proposed method quantitatively, the average CPU time for topology optimization of the displacement inverter by using the traditional method and the PDR method is compared with each other in this section. Table 3.4 shows the average CPU time of each step during one optimization iteration where **T** denotes the overall CPU time of each optimization iteration. Note that in case $\omega = 0$, the traditional optimization model is used.

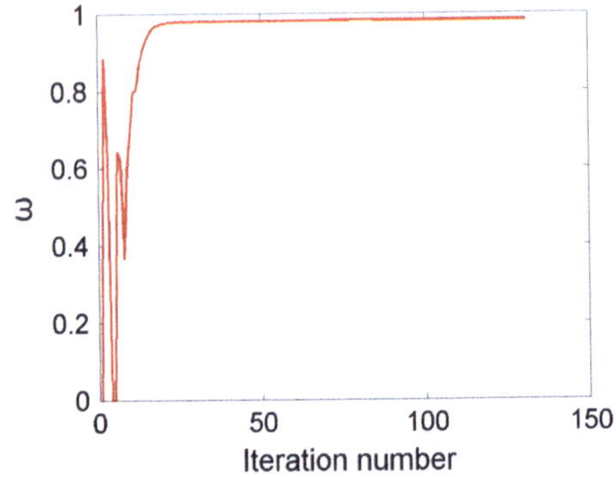

Fig. 3.12 The convergence history of ω

Table 3.4 The values of the average CPU times(s) in each iteration by using the traditional method and the PDR method

ω	Establish K	Compute U_1	Compute U_2	Sensitivity	Update **x**	**T**
0	4.2136	–	0.0914	0.0993	0.0247	4.4290
Eq. (3.19)	4.1988	0.0372	0.0901	0.1024	0.0242	4.4527

From Table 3.4 one can confirm that the establishment of **K** is the most CPU time consuming step during the topology optimization of the compliant mechanisms no matter which method is used.

Compared with the second and the third rows of the Table 3.4, one finds that by using the proposed method, the costing CPU time of each iteration is a little larger than the traditional method. This is mainly because an extra displacement field is needed to be calculated. However, it does not influence the computational efficiency very much since the establishment of **K** is only needed once, which means that the proposed method can yield hinge-free compliant mechanisms without losing the computational efficiency.

3.4 A Compliance-Based Method

3.4.1 Compliance Formulations

One widely used single loading condition for determining the stiffness is that the boundary of the applied load of the loading condition is considered to be fixed while a virtual load is applied at the output boundary [17]. Another loading condition for determining the stiffness that needs to be maximized is that only the input load is applied at the input boundary and an artificial spring is attached to the output boundary [23]. Nishiwaki [18] proposed two loading conditions for determining the stiffness. The first is that applying load at the input boundary and the output boundary is fixed, and the second is that the boundary of the applied load of the loading condition is considered to be fixed while a virtual load is applied at the output boundary.

Two types of mean compliances that can be incorporated into the optimization model are developed. Since two proposed loading conditions do not involve changing the input or output boundary into fixed and the mutual mean compliance as well can be determined based on the displacement field obtained from two proposed loading conditions, it can simplify the computational process of the finite element analysis.

Suppose that the original design domain **D** for topology design of compliant mechanisms is fixed at Γ_d and subjected to an input load F_{in} as shown in Fig. 3.13a. The first type of mean compliances is determined based on the case that an external unit force is only applied at the input port **i** while keeping the output port **o** as a free boundary(unfixed) as shown in Fig. 3.13b, denoted as C_{in}. The second type of mean compliance is determined based on the case that an external unit force is only applied at the output port **o** while keeping the input port **i** as a free boundary(unfixed) as shown in Fig. 3.13c, denoted as C_{out}.

The input compliance C_{in} and the output compliance C_{out} are respectively determined from the following equations:

$$C_{in} = \int_D f_{in} \cdot U_{C_{in}} d\Omega \tag{3.29}$$

$$C_{out} = \int_D f_{out} \cdot U_{C_{out}} d\Omega \tag{3.30}$$

where $U_{C_{in}}$ and $U_{C_{out}}$ are the displacement vectors due to f_{in} and f_{out}, respectively. Using the finite element method, the C_{in} and C_{out} can be simply written in the discrete form as

$$C_{in} = \mathbf{f}_{in}^T \mathbf{u}_{in} = \mathbf{u}_{in}^T \mathbf{K} \mathbf{u}_{in} \tag{3.31}$$

$$C_{out} = \mathbf{f}_{out}^T \mathbf{u}_{out} = \mathbf{u}_{out}^T \mathbf{K} \mathbf{u}_{out} \tag{3.32}$$

where $\mathbf{f}_{in}$ is a vector consisting of zero except for the input port position, where its value is f_{in} as shown in Fig. 3.13b and $\mathbf{u}_{in}$ is the displacement field due to $\mathbf{f}_{in}$, respectively. $\mathbf{f}_{out}$ is a vector consisting of zeros except for the output port position,

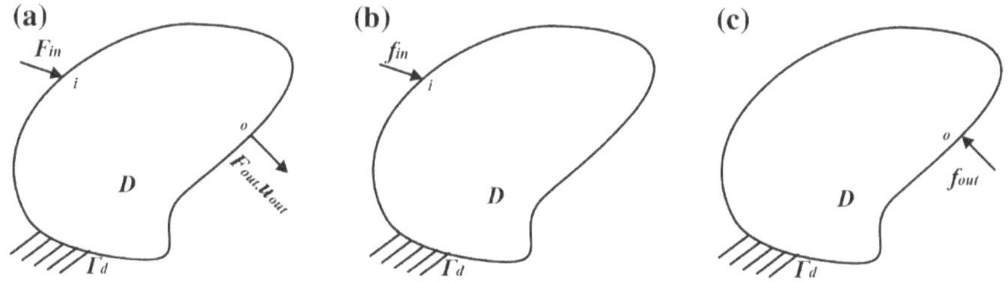

Fig. 3.13 Schematics for a compliant mechanisms design

where its value is f_{out} as shown in Fig. 3.13c and $\mathbf{u}_{out}$ is the displacement field due to $\mathbf{f}_{out}$, respectively. $\mathbf{K}$ is the stiffness matrix in the global level.

3.4.2 A GA-Based Formulation

One can choose the geometric advantage (GA) to quantify the performance of the compliant mechanisms. Therefore, incorporating the input and output compliances and using the weighted sum method will lead to a optimization model as follows

$$\min: \quad J = -GA + \alpha C_{in} + \beta C_{out} \tag{3.33}$$

$$\text{s. t.}: \quad Vol \leq Vol^{max} \tag{3.34}$$

where α and β are weighting factors. Note that the constraint on the input displacement is no longer needed due to the input compliance is taken into consideration.

As aforementioned, the choosing of the values of α and β is never an easy task. To analyze the statement further and demonstrate the influence of the weighting factors α and β, the displacement inverter problem is resolved by setting the values of α and β to five different combinations.

The optimization results for five different cases are sequentially shown in Table 3.5. The allowable material volume ratio is all set to be 25%. From Table 3.5, one can find that different α and β result in different optimizing results with specific GA. It should also be noted that although not all combinations of α and β can lead to distributed design as shown by Case 2 in Table 3.5, appropriate values of α and β do generate distributed results as shown by Case 3 in Table 3.5.

Case 1 in Table 3.5 shows that by setting α to be quite large and β to be 0, i.e., ignoring the influence of the second type of the mean compliance, a structure that only connects the fixed area with input port can be obtained. On the contrary, by setting α to be 0 and β to be quite large, i.e., ignoring the influence of the first type of the mean compliance, a structure that only connects the fixed area with output port can be obtained. Those are the results that have the structural configuration but can not accomplish the kinematic performance, i.e., the GA is almost 0, and the created mechanism is far beyond the widely recognized one.

When α and β are set to be small, a compliant mechanism that still has the de facto hinges can be obtained as indicated by Case 2 in Table 3.5. This kind of compliant mechanism mainly gains the motion from the deformation of the flexible hinges and is prone to suffer from the over stress problem *et al.* However, from this case, one can also realize that there is a new way to solve the disconnected problem in optimal design of compliant mechanisms proposed in [29]. By setting quite a small value to the weighting factor β of the second type of the mean compliance, the connectivity of the compliant mechanisms can be preserved during the optimization process.

As indicated by Case 2 and 4 in Table 3.5, which one is bigger between α and β has a certain impact on the result of the created mechanisms. By setting α too smaller than β as indicated by Case 4, one finds that a compliant mechanism with small GA can be obtained. This is more like a structure that exits snap-through behavior. As indicated by Case 3, a distributed compliant mechanism can be obtained by setting appropriate values of α and β. The created mechanism can satisfy the kinematic performance and have the structural configuration and therefor more suitable for the MEMS.

Table 3.5 Topology optimization of the displacement inverter using $Q1$ with different values of α and β

Initial design	Case	α	β	GA	Final design
	1	0.0100	0.0000	0	
	2	0.0040	0.0004	1.51	
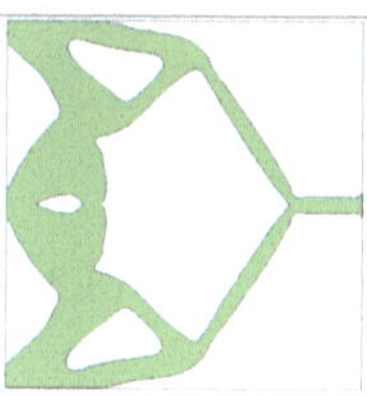	3	0.0020	0.0010	0.66	
	4	0.0004	0.0040	0.27	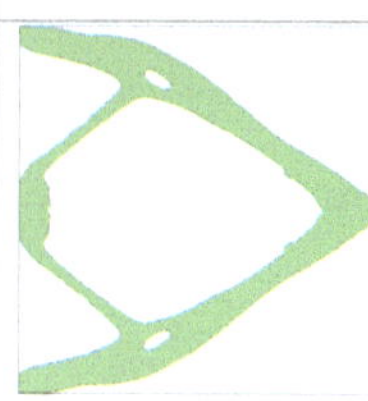
	5	0.0000	0.0100	0	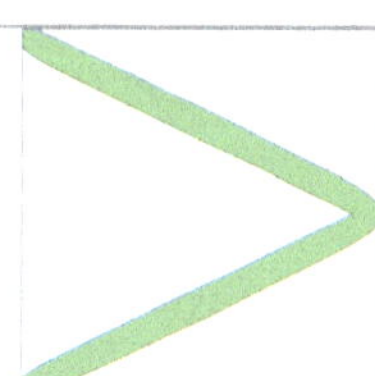

The above results show the importance of choosing α and β. The improper value of α or β either leads to compliant mechanisms with de facto hinges or results in compliant mechanisms with poor performance. Although in most cases we always can find out the appropriate values for α and β, naturally, this is not sufficient in many applications. And also, the value of the weighted factor depend on not only the material properties but also external conditions. In such cases a more efficient scheme for finding the appropriate values of the weighting factor would be necessary.

3.4.3 A Mutual Mean Compliance-Based Optimization Model

3.4.3.1 Mutual Mean Compliance

As shown in Fig. 3.14, suppose a linear elastic body occupying a 2D domain **D** and fixed at boundary Γ_d. Considering two cases which are Case (a) and Case (b). In Case (a), the design domain is subjected to a traction $\mathbf{t}_{in}$ at boundary $\Gamma_{\mathbf{t}_{in}}$ while in Case (b) the design domain is subjected to a traction $\mathbf{t}_{out}$ at boundary $\Gamma_{\mathbf{t}_{out}}$ as shown respectively in Fig. 3.14. Suppose that $\mathbf{t}_{in}$ is the applied input force and the corresponding displacement field is $\mathbf{u}_{in}$, and $\mathbf{t}_{out}$ is a unit dummy load and the corresponding displacement field is $\mathbf{u}_{out}$. Now the mutual mean compliance C_M can be used as a measure of the flexibility of the designed compliant mechanism and defined as:

$$C_M = L^{out}(\mathbf{u}_{in}) = a(\mathbf{u}_{out}, \mathbf{u}_{in}) \tag{3.35}$$

where

$$L^{out}(\mathbf{u}_{in}) = \int_{\Gamma} \mathbf{t}_{out}\mathbf{u}_{in} d\Gamma \tag{3.36}$$

$$a(\mathbf{u}_{out}, \mathbf{u}_{in}) = \int_{D} E_{ijkl}\varepsilon_{kl}(\mathbf{u}_{out})\varepsilon_{ij}(\mathbf{u}_{in}) d\Omega \tag{3.37}$$

where E_{ijkl} and ε are elasticity tensor and linearized strain tensor, respectively.

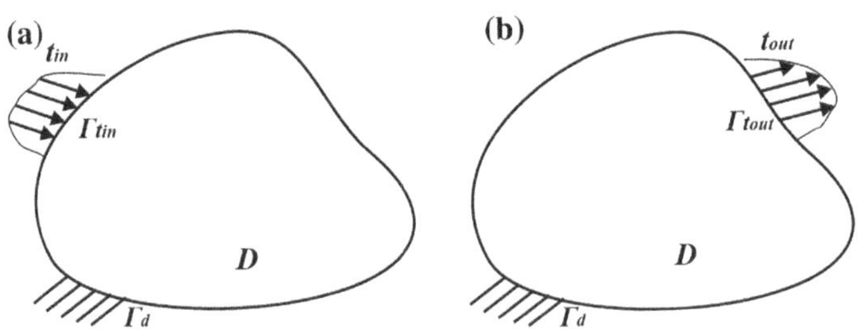

Fig. 3.14 Schematic for calculating mutual mean compliance

The mutual mean compliance C_M is interpreted as the deformation of boundary $\Gamma_{t_{out}}$ when $\mathbf{t}_{in}$ is applied at $\Gamma_{t_{in}}$. The larger the C_M is, the more flexible(compliant) the designed mechanism is. Therefore, to accomplish the requirement of flexibility of topology optimization of compliant mechanisms, C_M should be maximized.

Based on the above analysis, the mutual mean compliance C_M can be written as

$$C_M = F_{in}\mathbf{u}_{out}^T\mathbf{K}\mathbf{u}_{in} \tag{3.38}$$

Towards the optimization of compliant mechanisms, the created mechanisms should satisfy kinematic and stiffness requirements. The optimization problem for topology optimization of compliant mechanisms incorporating above ideas can be transferred into maximizing a monotonically increasing function $J(C_M, C_{in}, C_{out})$ of C_M, C_{in} and C_{out}.

Two widely used methods dealing with the multi-objective optimization problems are the weighting sum method and the ratio method which can be respectively written as follows

$$J = \varpi C_M - (1-\varpi)[\omega C_{in} + (1-\omega)C_{out}] \tag{3.39}$$

$$J = \frac{C_M}{\omega C_{in} + (1-\omega)C_{out}} \tag{3.40}$$

where ϖ and ω are the weighting factors.

The values of ϖ and ω used in Eq. (3.39) or Eq. (3.40) need to be pre-set. Constants are always employed through the whole optimization process [17] and different values will lead to different topologies. It is difficult to find suitable values mathematically. Furthermore, choosing the values of ϖ and ω in a global sense is rather difficult. Different design problems may need different weighting factors.

3.4.3.2 A Self-adjust Scheme for Setting the Weighting Factors

The optimization model $A0$ which is composed of objective function and the maximum material usage constraint therefore can be written as follows:

$$A0: \quad \min: \quad J = -C_M + \alpha C_{in} + \beta C_{out} \tag{3.41}$$

$$\text{s.t.}: \quad Vol \leq Vol_{max} \tag{3.42}$$

where α and β are weighting factors for C_{in} and C_{out}, respectively. Vol_{max} is the maximum allowable material usage.

A new scheme for updating the values of the weighting factors is developed. The underlying idea is to prevent the stiffness or the flexibility of a resulting compliant mechanism approaching too high, i.e., the proposed method can guarantee a well trade-off between flexibility and stiffness. Therefore, the de facto hinges can be successfully prevented.

If C_M is much larger than C_{in} and C_{out}, a very weak and fragile structure will be obtained. Conversely, if C_{in} and C_{out} are much larger than C_M, a very stiff structure will be obtained and it can not accomplish the kinematic requirement of the compliant mechanism. Therefore, a simple way to choose the values of the weighting factors α and β is that they can make the orders of magnitudes of the C_M, αC_{in} and βC_{out} at the same level.

Define a Lagrangian to convert the constrained optimization problem $A0$ into an unconstrained problem written as follows

$$L = -C_M + \alpha^k C_{in} + \beta^k C_{out} + \lambda(Vol + \chi^2 - Vol_{max}) \tag{3.43}$$

where χ is a slack variable to convert the inequality mass constraint into an equality one. α^k and β^k are weighting factors changing with each iteration k of the optimization algorithm and they are updated using

$$\alpha^{k+1} = \begin{cases} \frac{C_M^k}{C_{in}^k} & k \geq 1 \\ 0 & k = 0 \end{cases} \tag{3.44}$$

$$\beta^{k+1} = \begin{cases} \frac{C_M^k}{C_{out}^k} & k \geq 1 \\ 0 & k = 0 \end{cases} \tag{3.45}$$

where C_M^k, C_{in}^k and C_{out}^k denote the values of C_M, C_{in} and C_{out} in the kth iteration, respectively.

From Eqs. (3.43), (3.44) and (3.45), the following equation can be obtained:

$$\begin{aligned} & C_M^{k+1} : \alpha^{k+1} C_{in}^{k+1} : \beta^{k+1} C_{out}^{k+1} \\ & = (C_M^k + \Delta C_M) : (C_M^k + \frac{C_M^k}{C_{in}^k} \Delta C_{in}) : (C_M^k + \frac{C_M^k}{C_{out}^k} \Delta C_{out}) \end{aligned} \tag{3.46}$$

where ΔC_M, ΔC_{in} and ΔC_{out} are increments of C_M, C_{in} and C_{out} from iteration k to $k+1$ respectively.

Since compared to C_M, C_{in} and C_{out}, ΔC_M, ΔC_{in} and ΔC_{out} are small especially when the optimization process is close to the convergence, the following equation can be obtained:

$$C_M^{k+1} : \alpha C_{in}^{k+1} : \beta C_{out}^{k+1} \approx C_M^k : C_M^k : C_M^k = 1 : 1 : 1 \tag{3.47}$$

One can see that during iteration $k+1$, if C_M^k is larger than C_{in}^k or C_{out}^k, then $\alpha^{k+1} > 1$ or $\beta^{k+1} > 1$. This will increase the weight of C_{in} or C_{out} to make the created mechanism much stiffer. Conversely, if C_M^k is smaller than C_{in}^k or C_{out}^k, then $\alpha^{k+1} < 1$ or $\beta^{k+1} < 1$. This will decrease the weight of C_{in} or C_{out} to make the created mechanism more flexible.

From the above analysis one can see that for a n steps optimization problem, it has n objective functions. However, there is no need for manually operating the functions because the weighting factors can actually self-adjust based on the information obtained from the previous iteration.

In the first iteration, e.g., $k = 1$, since α and β are set to be 0, the optimization problem $A0$ is actually reduced to

$$\text{min}: \quad J = -C_M \tag{3.48}$$

$$\text{s. t.}: \quad Vol \leq Vol_{\max} \tag{3.49}$$

This will lead to a very large u_{out} meanwhile the very small C_{in} and C_{out}. Therefore, a very weak structure can be obtained and it can not perform kinematic outputs. However, after the first iteration, α and β will become very large to enhance the weighting of C_{in} and C_{out}. Therefore, a well-posed design still can be obtained.

3.4.3.3 Alternative Formulations

Based on the weighting factors setting scheme, it is easy to note that there are two alternative functions. The first of which can be stated as follows:

$$A1: \quad \text{min}: \quad J = -\alpha_{A1} C_M + C_{in} + \beta_{A1} C_{out} \tag{3.50}$$

$$\text{s. t.}: \quad Vol \leq Vol_{max} \tag{3.51}$$

where α_{A1} and β_{A1} are the weighting factors for C_M and C_{out}, respectively. The scheme for updating the α_{A1} and β_{A1} can be stated as follows:

$$\alpha_{A1}^{k+1} = \begin{cases} \frac{C_{in}^k}{C_M^k} & k \geq 1 \\ 0 & k = 0 \end{cases} \tag{3.52}$$

$$\beta_{A1}^{k+1} = \begin{cases} \frac{C_{in}^k}{C_{out}^k} & k \geq 1 \\ 0 & k = 0 \end{cases} \tag{3.53}$$

where C_M^k, C_{in}^k and C_{out}^k denote the values of C_M, C_{in} and C_{out} in the k-th iteration, respectively. α_{A1}^{k+1} and β_{A1}^{k+1} are the values of α_{A1} and β_{A1} in the $(k+1)$-th step, respectively. One can see that the following results can be easily obtained by following the way proposed in Eq. (3.46):

$$\alpha_{A1} C_M^{k+1} : C_{in}^{k+1} : \beta_{A1} C_{out}^{k+1} \approx C_{in}^k : C_{in}^k : C_{in}^k = 1 : 1 : 1 \tag{3.54}$$

The second alternative function can be expressed as follows:

$$A2: \quad \min: \quad J = -\alpha_{A2}C_M + \beta_{A2}C_{in} + C_{out} \tag{3.55}$$

$$\text{s. t.}: \quad Vol \leq Vol_{max} \tag{3.56}$$

where α_{A2} and β_{A2} are the weighting factors for C_M and C_{in}, respectively. The scheme for updating the α_{A2} and β_{A2} can be stated as follows:

$$\alpha_{A2}^{k+1} = \begin{cases} \frac{C_{out}^k}{C_M^k} & k \geq 1 \\ 0 & k = 0 \end{cases} \tag{3.57}$$

$$\beta_{A2}^{k+1} = \begin{cases} \frac{C_{out}^k}{C_{in}^k} & k \geq 1 \\ 0 & k = 0 \end{cases} \tag{3.58}$$

where C_M^k, C_{in}^k and C_{out}^k denote the values of C_M, C_{in} and C_{out} in the k-th iteration, respectively. α_{A2}^{k+1} and β_{A2}^{k+1} are the values of α_{A2} and β_{A2} in the $(k+1)$-th step, respectively. Similarly, the following results can be easily obtained by following the way proposed in Eq. (3.46):

$$\alpha_{A2}C_M^{k+1} : \beta_{A2}C_{in}^{k+1} : C_{out}^{k+1} \approx C_{out}^k : C_{out}^k : C_{out}^k = 1 : 1 : 1 \tag{3.59}$$

The method is developed based on a new design model and self-adjust weighting factors choosing scheme without special requirements. Therefore, the implementation of the proposed method is extremely easy since no extra constrains need to be considered. Although the weighting factors are remaining changed during the optimization process, their derivatives with respect to the design variables are 0. Therefor, the sensitivities of the proposed objective function are also cheap to compute.

3.4.3.4 Numerical Examples

In order to demonstrate the validity of the developed formulations, the displacement inverter design problem is resolved using the level set method. The maximal material usage is restricted to 20%. The design domain is discretized with 80×80 quadrilateral elements. Note that due to the symmetry, only the lower half part of the design domain is taken into consideration for elastic analysis.

The optimization process was run for 200 iterations. The final topology and the corresponding local energy density are shown in Fig. 3.15 in which the final design obtained using the conventional optimization model is also shown.

Comparison Study It should be noted that the present formulations can ensure elimination of the de facto hinges. Figure 3.15b shows that a continuous, hinge-free compliant displacement inverter can be obtained by using $A0$ without using any extra constrains. The broken parts that are shown inside the circles in Fig. 3.15a are the de

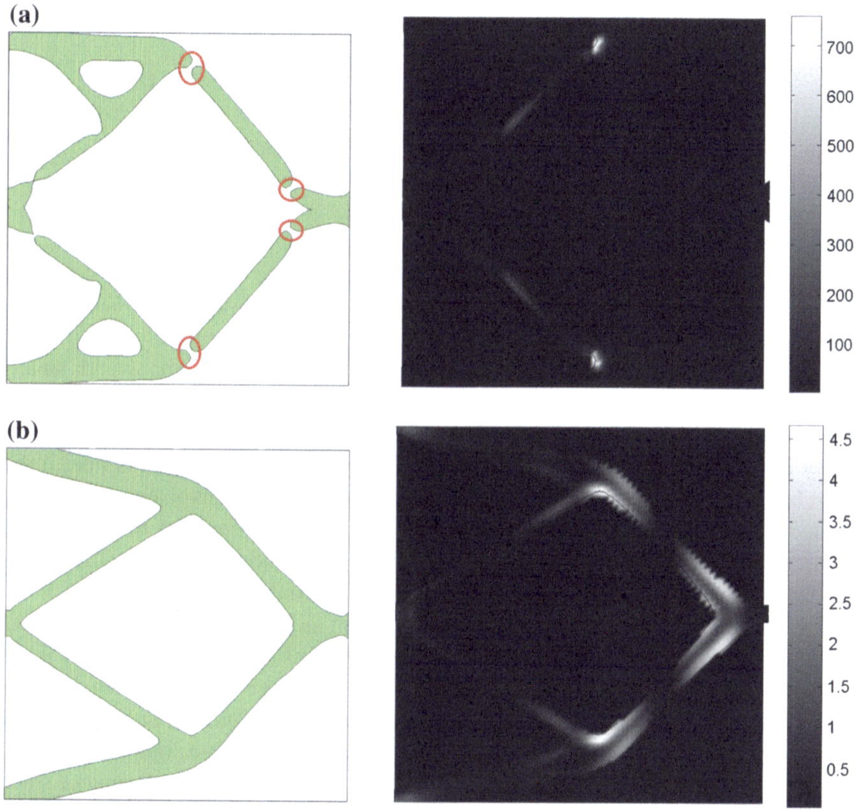

Fig. 3.15 The final design and the corresponding local energy density of the displacement inverter: **a** The hinged design; **b** Hinge-free design

facto hinge regions. The flexibility of the design is mainly concentrated in the hinged regions which can be seen from its local energy density distribution.

When the hinge-free displacement inverter is loaded, nearly all parts of the mechanism are contributing to the deformation as shown in Fig. 3.15b. When a lumped displacement inverter as shown in Fig. 3.15 is loaded, just material around the hinge regions (see Fig. 3.15a) will contribute to the deflection at the output port. This makes the lumped compliant mechanisms suffer hinge stress concentration. From this point of view, the hinge-free compliant mechanisms are more preferable to the lumped compliant mechanisms in the sense of decreasing stress concentration and possible fatigue breakage.

The set of computation results presented in Table 3.6 intends to make a quantitative comparison between the hinged displacement inverter and the hinge-free design obtained in this work. The numerical results of the *Hinged* row are obtained based on the device as shown in Fig. 3.15a and the numerical results of the *hinge-free* row are obtained based on the device as shown in Fig. 3.15b. They have the same material

Table 3.6 Characteristics comparison between hinged displacement inverter and the hinge-free design

Type	Input load	Input dis	Output dis	Maximum stress
Hinged	100	42.1210	−38.9983	765.1055
Hinge-free	100	11.0198	−5.8856	7.033

usage constraint, the same material properties and the same input load. Note that the maximum Von-Mises stress of the hinged inverter approach very high because the elements in the hinged areas have very small stiffness and very large deformation. The high stress condition can be improved by replacing the de facto hinges with small-length beams. However, the obtained compliant mechanisms using the replacement method are still suffering the stress concentration problem.

By eliminating the de facto hinges, the maximum Von-Mises stress due to the input load can be tremendously reduced. Compared the input displacements of two studied mechanisms due to the same input load, one can see that the hinge-free inverter obtained in this work is stiffer. And also the introduced mean compliances clearly have direct impact on the magnitude of the output displacements.

For obtaining a continuous, hinge-free complaint mechanism that has higher flexibility, one can impose much more restrictive material usage constraint values. Since the resulting compliant mechanisms do not suffer the hinge problem, a small material usage constraint can be used from achieving higher flexibility. Further, from the results shown in Table 3.6 one can see that the obtained hinge-free compliant mechanisms can bear a large input load for achieving a higher kinematic performance since the stress concentration is no longer an issue.

Figure 3.16a–f represent some intermediate designs in their half form. Topology change is mainly concentrated at the first 120 iterations. The proposed method can prevent the de facto hinges in the resulting compliant mechanisms not only in the

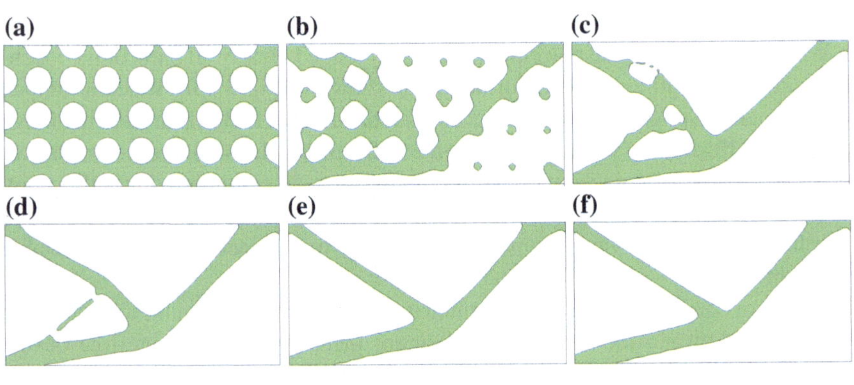

Fig. 3.16 The intermediate designs of the displacement inverter: **a** step 1; **b** step 30; **c** step 75; **d** step 120; **e** step 150; and **f** step 180

final design but also during the optimization process. With the evolving the level set function, the topology change such as merging can be naturally obtained. Note that, during the optimization, new holes can not be generated. For using the proposed formulations based on a conventional level set method, a certain number of holes should be preseted in the initial design. However, the authors believe that reasonable designs also can be obtained by using the proposed formulations incorporating with other level set method such as the parameterization level set method.

Figure 3.17 shows the convergent curves of the C_M, volume ratio, two types of mean compliances and the lagrange objective function ranging from 0 to 200 iterations. When the C_M increases too high due to the violation of the volume constraint, the C_M will change in an opposite direction due to the violation of two mean compliances to avoid generating a disconnected structure due to high flexibility. Figure 3.17 shows that the optimization is almost completed at iteration 100. However, from Fig. 3.17 one can see that the last 100 iterations are still necessary to ensure a uniform distribution of the material.

The multi-objective function proposed in this study actually provides two kinds of velocity fields. The first kind is produced by C_M which makes the created mechanism

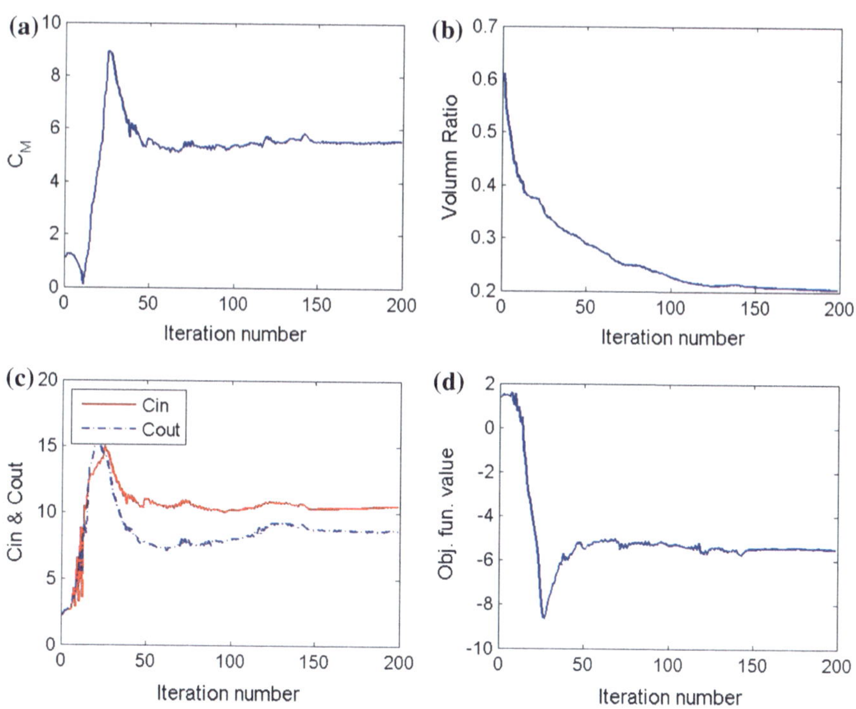

Fig. 3.17 Convergence history of the displacement inverter: **a** C_M; **b** Volume ratio; **c** $C_{in}\&C_{out}$; and **d** Objective function

a compliant one. The second kind is produced by the weighting sum of the C_{in} and C_{out} which make the created compliant mechanism free of de facto hinges.

The main difference among the multi-objective scheme proposed in this study and other multiple objective functions is that the weighting factors of the objectives does not need to be pre-set and they can self-adjust during the optimization process. Figure 3.18 shows the weighting factors curves of two types of mean compliances. At the first 80 iterations, due to the topology change, α and β change by a relatively large amount. And after 80 iterations, α and β basically remain unchanged. One can see that during the whole optimization process, due to the change of weighting factors, αC_{in}, βC_{out} and C_M are remaining in the same level.

Results Obtained form A1 and A2 There are three possible optimization models for topology optimization of hinge-free compliant mechanisms. The above results have proved the validity of the optimization model *A*0. In order to examine the validity of the optimization models *A*1 and *A*2, the displacement inverter is resolved. The design domain is also dicretized by using 80×80 elements for the elastic analysis. The maximum material usage constraint is set to be 20%.

The final designs of the displacement inverter obtained by using *A*1 and *A*2 are shown in Fig. 3.19a and b, respectively. Nearly identical results can be obtained with the outcome obtained by using *A*0. The results confirm the proposed self-adjust weighting factors scheme for designing the hinge-free compliant mechanisms.

There are three objectives in the *A*0(or *A*1, *A*2) that need to be minimized or maximized which are C_M, C_{in} and C_{out}. The main difference among the *A*0, *A*1 and *A*2 is that the reference objective is differently chosen. For instance, in the *A*0, C_M is chosen as the reference objective while in the *A*1 the C_{in} is chosen. Therefore, during the optimization process, the values of the weighting factors in the *A*0 are different from the values in the *A*1. Since the same results can be obtained by using either three of the optimization models, the following numerical results through the article are also obtained based on the *A*0.

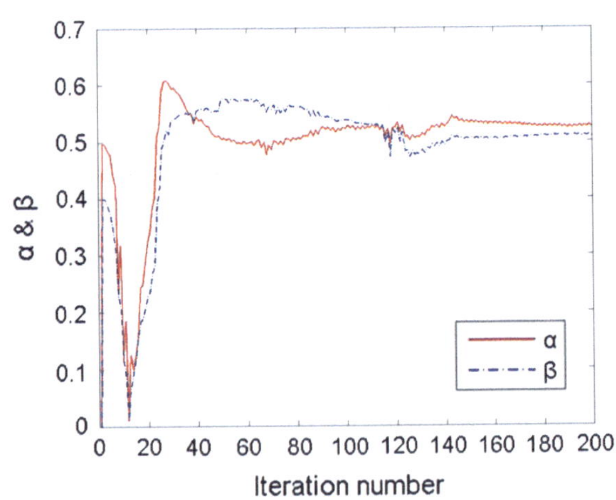

Fig. 3.18 Weighting factors curve of the displacement inverter

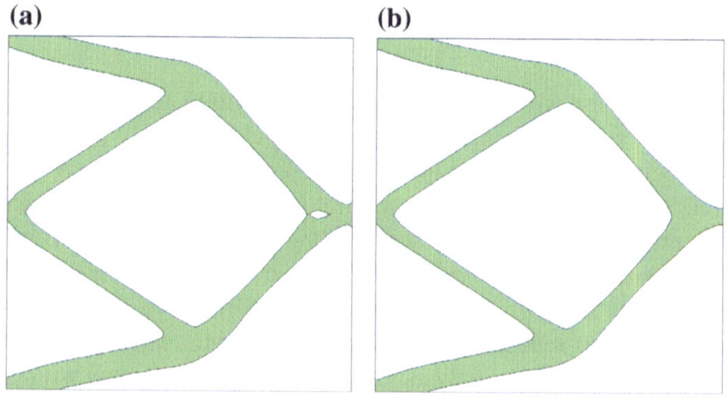

Fig. 3.19 Final designs of displacement inverter using different optimization model: **a** $A1$, and **b** $A2$.

Design with Different Mesh Refinement Since the proposed method is independent of the underlying finite element mesh refinement, the optimal design will also be independent of the mesh refinement in general. Thus, the result of the optimization problem should be not changed if the finite element mesh is refined. To study mesh independence, three cases are studied.

The design domain is discretized by using $60 \times 60 = 3600$, $120 \times 120 = 14400$ and $160 \times 160 = 25600$ elements, respectively. The maximal material usage is restricted to 20%. The optimum material distributions are shown in Fig. 3.20, respectively.

Note that de facto hinges are not occurring in all the designs, even through a slightly bad discretization is used (60×60). The consistent designs can be obtained with the different meshing refinements. For the finite element analysis, a fine discretization is surely capable for capturing the spatial partial derivative and the elastic analysis. The examples illustrate that the proposed method dose not suffer the mesh-dependent problem.

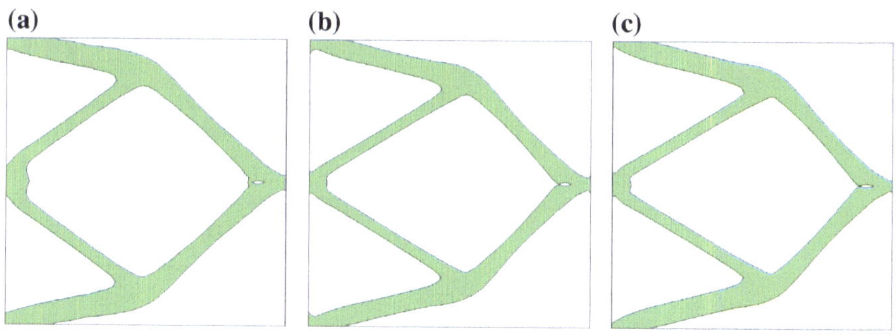

Fig. 3.20 Final designs of displacement inverter with different mesh refinement: **a** 60×60, **b** 120×120, and **c** 160×160.

3.5 Compliant Mechanisms with Multiple Outputs

Compliant mechanisms with multiple inputs and multiple outputs are widely used in the fields of micro-positioning and micro-manipulation. Only a small amount of papers which focus on topology optimization of compliant mechanisms with multiple degree of freedom. For instance, based on a multi-criteria optimization procedure [7], Frecher et al. proposed two methods for handling the multiple output requirements, i.e., a combined virtual load method and a weighted sum of objectives method based on the ground structure approach. Saxena [23] performed a genetic algorithm for synthesizing compliant mechanisms for desired multiple output displacements by using geometrically nonlinear analysis. Combined the strain energy with the mutual potential energy, Zhang [30] developed a multiple objective function for topology design of multiple inputs and outputs compliant mechanisms.

If one consider the design with a compliant mechanism with multiple outputs, the design domain can be illustrated as in Fig. 3.21. The problem is posed in terms of an input load and multiple specified output deflections.

The goal of the optimization problem can be grouped into two parts. The first part is to maximize each desired displacement Δ^i_{out}, $i = 1, 2, \ldots, n$ in the specified direction due to the actual input load F_{in} which can be defined by maximizing a monotonically increasing function of Δ^i_{out}, i.e., $f(\boldsymbol{\Delta}_{\mathbf{out}})$, where $\boldsymbol{\Delta}_{\mathbf{out}}$ is the vector of output displacements defined as:

$$\boldsymbol{\Delta}_{\mathbf{out}} = [\Delta^1_{out}\ \Delta^2_{out}\ \cdots\ \Delta^n_{out}]^T \tag{3.60}$$

The second part is to make the created compliant mechanisms *hinge-free* and can be accomplished by taking two types of mean compliances into account.

The first one is called the input mean compliance C_{in} which has been defined before. It is the case that only a load applied at the input port of the design domain which can be formulated as

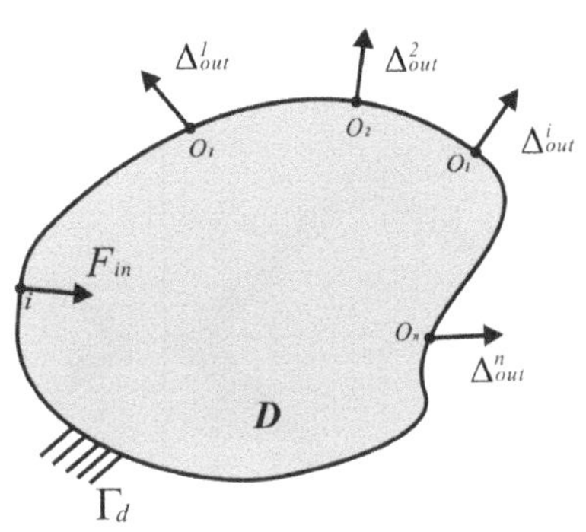

Fig. 3.21 The design domain of compliant mechanisms with multiple outputs

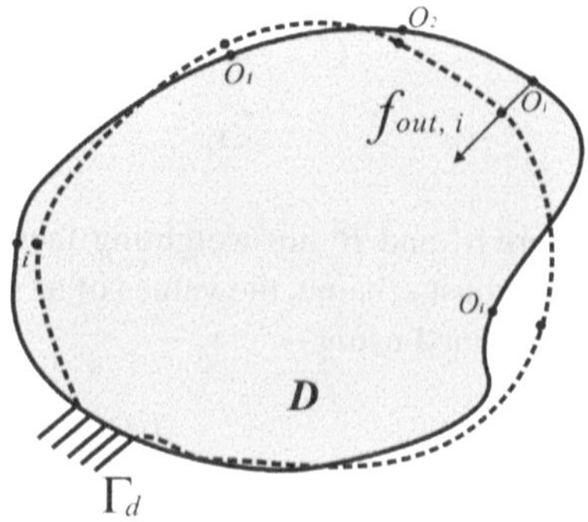

Fig. 3.22 The boundary condition for determining the output mean compliance $C_{out,i}$

$$C_{in} = \int_D f_{in} \cdot U_{in} d\Omega \tag{3.61}$$

where U_{in} is the displacement vector due to the virtual load f_{in}.

The second type of the mean compliance can be called the output mean compliance C^i_{out}, $(i = 1, 2, \ldots, n)$ which can be defined as the case that only a unit load applied at the output port. As shown in Fig. 3.22, C^i_{out}, $(i = 1, 2, \ldots, n)$ can be formulated as

$$C^i_{out} = \int_D f_{out,i} \cdot U^i_{out} d\Omega;\ i = 1, 2, \ldots, n. \tag{3.62}$$

where U^i_{out} is the displacement vector due to the virtual load $f_{out,i}$.

An increasing function of C_{in} and C^i_{out}, i.e., $g(\mathbf{C})$, can be used in the objective function to prevent generating flexible hinges during the optimization process, where $\mathbf{C}$ is the vector of two types of mean compliances which can be defined as:

$$\mathbf{C} = [C_{in}\ C^1_{out}\ C^2_{out}\ \cdot\ C^n_{out}]^T \tag{3.63}$$

3.5.1 *Optimization Formulation*

A multi-criteria objective problem is considered. Similarly, for topology optimization of hinge-free compliant mechanisms with multiple outputs, one can formulate it as the weighted sum of the maximizing $f(\Delta_{\mathbf{out}})$ and minimizing $g(\mathbf{C})$. In order to overcome the difficult of selecting the appropriate weighting factors, one can adopted the aforementioned self-adjust scheme.

Therefore, the optimization model for topology optimization of compliant mechanisms with multiple outputs can be formulated as follows

$$
\begin{aligned}
Q_{mo}\quad \min: &\quad J = -\sum_{i=1}^{n} \alpha^i \Delta_{out}^i + C_{in} + \sum_{i=1}^{n} \beta^i C_{out}^i \\
\text{s.t.}: &\quad Vol \leq Vol_{max}
\end{aligned}
\tag{3.64}
$$

where α^i and β^i are weighting factors of Δ_{out}^i and C_{out}^i, respectively. In using the self-adjust scheme, the values of α^i and β^i are updated with iterations, which can be determined using

$$
(\alpha^i)^{k+1} = \begin{cases} |\frac{C_{in}^k}{(\Delta_{out}^i)^k}| & \text{if } k > 0 \\ 0 & \text{if } k = 0 \end{cases} \tag{3.65}
$$

$$
(\beta^i)^{k+1} = \begin{cases} |\frac{C_{in}^k}{(C_{out}^i)^k}| & \text{if } k > 0 \\ 0 & \text{if } k = 0 \end{cases} \tag{3.66}
$$

where $(\Delta_{out}^i)^k$, C_{in}^k and $(C_{out}^i)^k$ are the values of Δ_{out}^i, C_{in} and C_{out}^i in the iteration k, respectively. Specially, in the first iteration, the values of α^i and β^i can be simply set to 0.

3.5.2 Validation

In order to demonstrate the validity of the formulation Q_{mo} on designing hinge-free compliant mechanisms with multiple outputs, a displacement redirector is considered in this section. The design problem is shown in Fig. 3.23, where the input force F_{in} is applied at the middle of the left side to cause two outputs deflections Δ_{out}^1 and Δ_{out}^2 at the points o_1 and o_2, respectively.

The whole design domain, including the void area, is discretized using 80×80 finite elements for elastic analysis. The maximal material usage is restricted to be 30%. The optimization process was run for 32 iterations.

The level set method is adopted to solve the optimization problem. The initial, intermediate and final design of the compliant displacement redirector mechanisms

Fig. 3.23 The design domain of the compliant displacement redirector mechanisms with two outputs

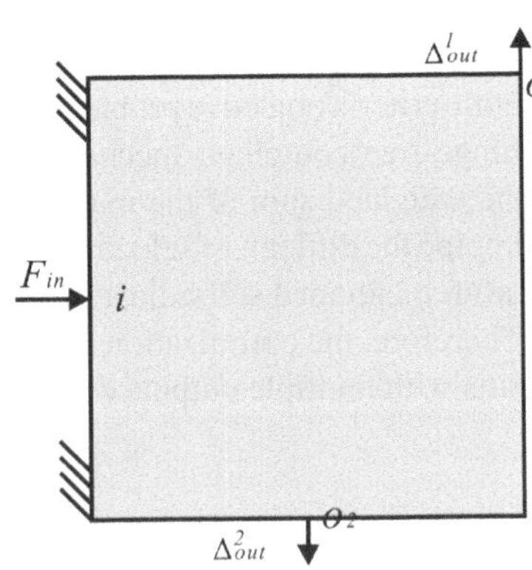

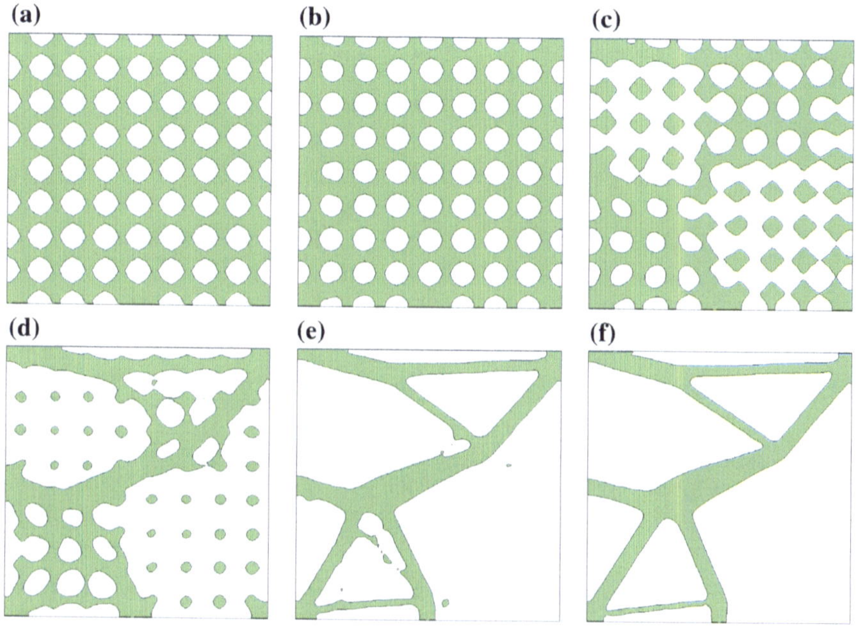

Fig. 3.24 The initial, intermediate and final design of the compliant displacement redirector mechanisms with two outputs: **a** initial configuration; **b** step 1; **c** step 5; **d** step 10; **e** step 20; **f** final configuration

are shown in Fig. 3.24. The de facto hinges are completely eliminated. One can see that the proposed method in this paper can prevent generating flexible hinges not only in the final design but also during the optimization process. Therefore, the created mechanism no longer suffers the high stress concentration. And this make the created mechanisms more suitable for real world applications. The introduced mean compliances clearly have direct impact on the magnitude of the output displacements. However, the introduced mean compliances also can decrease the displacement due to the F_{in}, i.e., the stiffness of the created mechanisms is increased. To ensure a the maximum output displacement, a large input force can be used for the hinge-free compliant mechanisms.

Figure 3.25 shows the convergence histories of the Δ^1_{out}, Δ^2_{out} and volume ratio during the optimization process. Figure 3.26 shows the The convergence histories of C_{in}, C^1_{out} and C^2_{out} whereas Fig. 3.27. Oscillations occur at the initial 10 iterations. After α^i and β^i achieve a more stable level, the objective function uses 20 iterations to converge to a minimum.

In order to illustrate the effect of the direction of the output deflection requirements on the final design. By varying the direction of output deflection requirements at the point o_2, two extra cases are studied and the design domain is shown as in Fig. 3.28a and b, respectively.

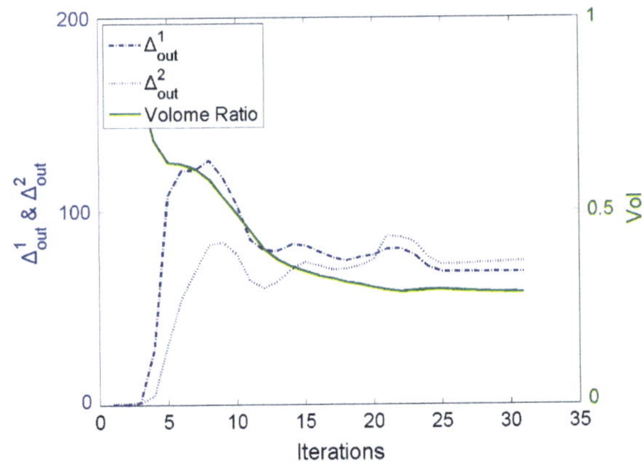

Fig. 3.25 The convergence histories of Δ^1_{out}, Δ^2_{out} and volume ratio

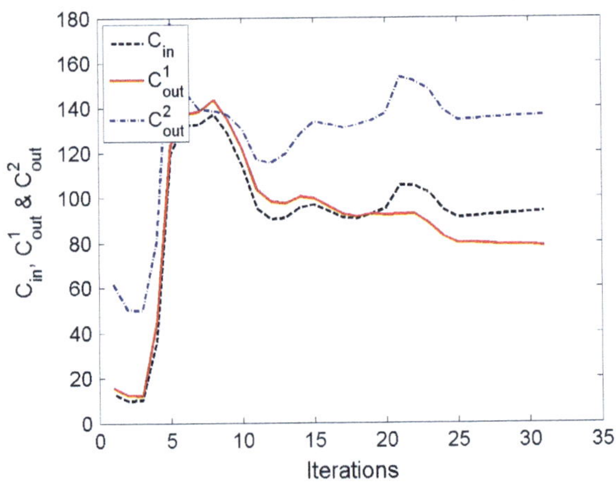

Fig. 3.26 The convergence histories of C_{in}, C^1_{out} and C^2_{out}

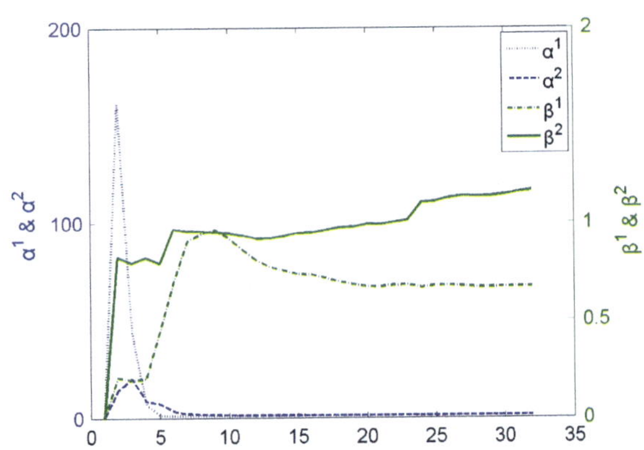

Fig. 3.27 The convergence histories of α^1, α^2, β^1 and β^2

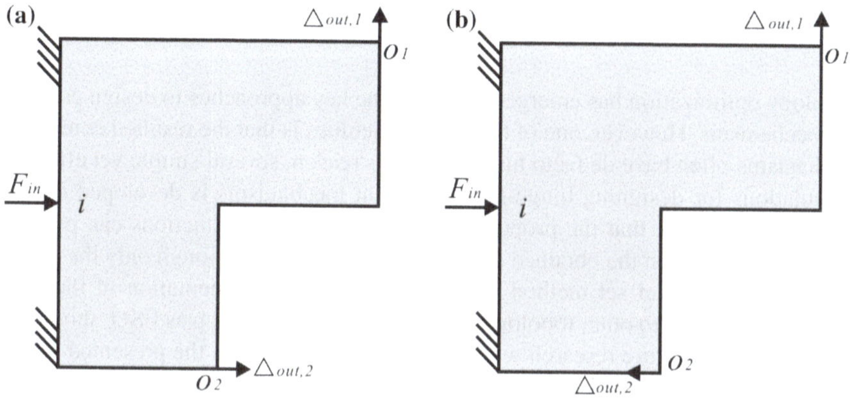

Fig. 3.28 The design domains of the two outputs mechanism with different output deflection requirements: **a** Case 1; **b** Case 2

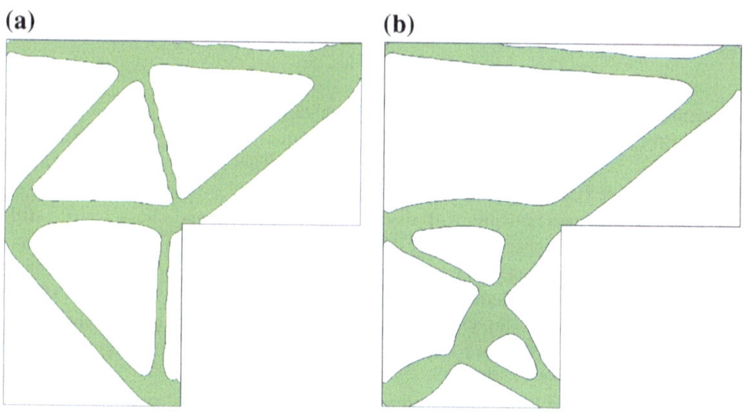

Fig. 3.29 The corresponding final design of the two outputs mechanism: **a** Case 1; **b** Case 2

The corresponding topology results are shown as in Fig. 3.29a and b, respectively. The flexible hinges are completely eliminated for both of the studied cases.

As indicated by those two cases, the formulation Q_{mo} can only ensure the component of the output displacement along the desired direction is maximized. As shown in Fig. 3.29b, the vertical displacements at point o_1 and o_2 are bigger than the horizontal, i.e., the desired displacements. This may can be handled by adding extra constraints.

3.6 Conclusions

Topology optimization has emerged as one of the key approaches to design compliant mechanisms. However, one of the main difficulties is that the resulted compliant mechanisms often have de facto hinges. For this reason, several simple yet efficient formulations for designing hinge-free compliant mechanisms is developed in this chapter. It is shown that the proposed augmented objective functions can prevent the de facto hinges in the obtained compliant mechanisms. Although only the SIMP method and the level set method are considered, the implementation of the proposed formulations to other topology optimization methods, such as ESO, should be straightforward. Future research will investigate the validity of the presented methods for designing both 2D and three-dimensional (3D) compliant mechanisms with large displacement.

References

1. Ananthasuresh, G., Kota, S., Kikuchi, N.: Strategies for systematic synthesis of compliant mems. In: Proceedings of the 1994 ASME Winter Annual Meeting, pp. 677–686 (1994)
2. Bendsoe, M.P., Sigmund, O.: Topology Optimization: Theory, Methods, and Applications. Springer Science & Business Media (2013)
3. Chen, S.: Compliant mechanisms with distributed compliance and characteristic stiffness: a level set method. The Chinese University of Hong Kong, PR China, Ph.D. Dissertation (2007)
4. Chen, S.K., Gonella, S., Chen, W., Liu, W.K.: A level set approach for optimal design of smart energy harvesters. Comput. Methods Appl. Mech. Eng. **199**(37–40), 2532–2543 (2010)
5. Deepak, S.R., Dinesh, M., Sahu, D.K., Ananthasuresh, G.: A comparative study of the formulations and benchmark problems for the topology optimization of compliant mechanisms. J. Mech. Robot. **1**(1), 011,003 (2009)
6. Duysinx, P., Bendsøe, M.P.: Topology optimization of continuum structures with local stress constraints. Int. J. Numer. Meth. Eng. **43**(8), 1453–1478 (1998)
7. Frecker, M., Kikuchi, N., Kota, S.: Topology optimization of compliant mechanisms with multiple outputs. Structural Optimization **17**(4), 269–278 (1999)
8. Haber, R., Jog, C., Bendsøe, M.P.: A new approach to variable-topology shape design using a constraint on perimeter. Structural Optimization **11**(1–2), 1–12 (1996)
9. Hetrick, J.A.: An energy efficiency approach for unified topological and dimensional synthesis of compliant mechanisms. Ph.D. thesis, University of Michigan (1999)
10. Howell, L.L.: Compliant Mechanisms. Wiley (2001)
11. Joo, J., Kota, S., Kikuchi, N.: Topological synthesis of compliant mechanisms using linear beam elements. Mech. Structures Mach. **28**(4), 245–280 (2000)
12. Kim, J.E., Kim, Y.Y., Min, S.: A note on hinge-free topology design using the special triangulation of design elements. Int. J. Numerical Methods Biomed. Eng. **21**(12), 701–710 (2005)
13. Kota, S., Joo, J., Li, Z., Rodgers, S.M., Sniegowski, J.: Design of compliant mechanisms: applications to mems. Analog Integr. Circ. Sig. Process **29**(1–2), 7–15 (2001)
14. Lau, G., Du, H., Lim, M.: Use of functional specifications as objective functions in topological optimization of compliant mechanism. Comput. Methods Appl. Mech. Eng. **190**(34), 4421–4433 (2001)
15. Lee, E., Gea, H.C.: A strain based topology optimization method for compliant mechanism design. Structural Multidisciplinary Optimization **49**(2), 199–207 (2014)

16. Luo, J., Luo, Z., Chen, L., Tong, L., Wang, M.Y.: A semi-implicit level set method for structural shape and topology optimization. J. Comput. Phys. **227**(11), 5561–5581 (2008)
17. Luo, Z., Chen, L., Yang, J., Zhang, Y., Abdel-Malek, K.: Compliant mechanism design using multi-objective topology optimization scheme of continuum structures. Structural Multidisciplinary Optimization **30**(2), 142–154 (2005)
18. Nishiwaki, S., Min, S., Yoo, J., Kikuchi, N.: Optimal structural design considering flexibility. Comput. Methods Appl. Mech. Eng. **190**(34), 4457–4504 (2001)
19. Petersson, J., Sigmund, O.: Slope constrained topology optimization. Int. J. Numer. Meth. Eng. **41**(8), 1417–1434 (1998)
20. Poulsen, T.A.: A new scheme for imposing a minimum length scale in topology optimization. Int. J. Numer. Meth. Eng. **57**(6), 741–760 (2003)
21. Rahmatalla, S., Swan, C.C.: Sparse monolithic compliant mechanisms using continuum structural topology optimization. Int. J. Numer. Meth. Eng. **62**(12), 1579–1605 (2005)
22. Rozvany, G.I.: A critical review of established methods of structural topology optimization. Structural Multidisciplinary Optimization **37**(3), 217–237 (2009)
23. Saxena, A., Ananthasuresh, G.: An optimality criteria approach for the topology synthesis of compliant mechanisms. In: Proceedings of DETC98, ASME Design, Engineering Technical Conference (1998)
24. Sigmund, O.: On the design of compliant mechanisms using topology optimization. Mech. Structures Mach. **25**(4), 493–524 (1997)
25. Sigmund, O.: Design of multiphysics actuators using topology optimization-part I: one-material structures. Comput. Methods Appl. Mech. Eng. **190**(49), 6577–6604 (2001)
26. Sigmund, O.: Morphology-based black and white filters for topology optimization. Structural Multidisciplinary Optimization **33**(4–5), 401–424 (2007)
27. Wang, M.Y.: A kinetoelastic formulation of compliant mechanism optimization. J. Mech. Robot. **1**(2), 021,011 (2009)
28. Wang, M.Y., Chen, S.: Compliant mechanism optimization: analysis and design with intrinsic characteristic stiffness. Mech. Based Des. Structures Mach. **37**(2), 183–200 (2009)
29. Yamada, T., Izui, K., Nishiwaki, S.: A level set-based topology optimization method for maximizing thermal diffusivity in problems including design-dependent effects. J. Mech. Des. **133**(3), 031,011 (2011)
30. Zhang, X., Chen, Y.: Topology optimization of multiple inputs and outputs compliant mechanism with coupling terms control. Jixie Gongcheng Xuebao (Chin. J. Mech. Eng. **42**(3), 162–165 (2006)
31. Zhou, H.: Topology optimization of compliant mechanisms using hybrid discretization model. J. Mech. Des. **132**(11), 111,003 (2010)
32. Zhu, B., Zhang, X.: A new level set method for topology optimization of distributed compliant mechanisms. Int. J. Numer. Meth. Eng. **91**(8), 843–871 (2012)
33. Zhu, B., Zhang, X., Wang, N.: Topology optimization of hinge-free compliant mechanisms with multiple outputs using level set method. Structural Multidisciplinary Optimization **47**(5), 659–672 (2013)

Chapter 4
Topology Optimization of Compliant Parallel Mechanisms

Abstract This chapter introduces a new conceptual design method, the Jacobian-based topology optimization method, for compliant parallel mechanisms (CPMs). CPMs are usually synthesized using the rigid-body replacement approach, which means that mechanisms of this type rely on existing rigid-body mechanisms and lack a synthesis method that considers topology performance. To this end, the Jacobian-based topology optimization method tries to extend the idea of topology optimization to the synthesis of CPMs. Unlike traditional topology optimization methods, this method regards the separate compliant limbs of CPMs as design domains and is based on different analysis methods and problem formulations. The problem of optimizing the topology of a CPM is equivalent to finding the optimal compliant limbs of CPMs across multiple design domains. By optimizing the Jacobian matrix, which represents the input and output stiffness of the whole mechanism, the functionality and performance can be improved simultaneously.

4.1 Introduction

As a combination of parallel and compliant mechanisms, typical compliant parallel mechanisms (CPMs) are parallel mechanisms employing flexure hinges at their joints. Consequently, CPMs have the advantages of both parallel and compliant mechanisms in terms of high stiffness, high load carrying capacity, high precision, etc. The potential applications of CPMs are precision engineering situations that require micro- and nano-positioning techniques, such as scanning probe microscopy [35], cell microinjection [30], and precision alignment [20].

This chapter focuses on the conceptual design of CPMs. Currently, there are two main conceptual design approaches for compliant mechanisms, the rigid-body replacement approach [11, 29] and the topology optimization approach [3, 46].

CPMs are usually synthesized using the rigid-body replacement approach. This approach seldom creates new topologies for CPMs in response to design problems, but begins with the existing topologies of rigid-body mechanisms and replaces their kinematic joints with flexural hinges. See Fig. 4.1 for an example. Therefore, the majority of the CPMs proposed in the literature share the topologies of their

X. Zhang and B. Zhu, *Topology Optimization of Compliant Mechanisms*,
https://doi.org/10.1007/978-981-13-0432-3_4

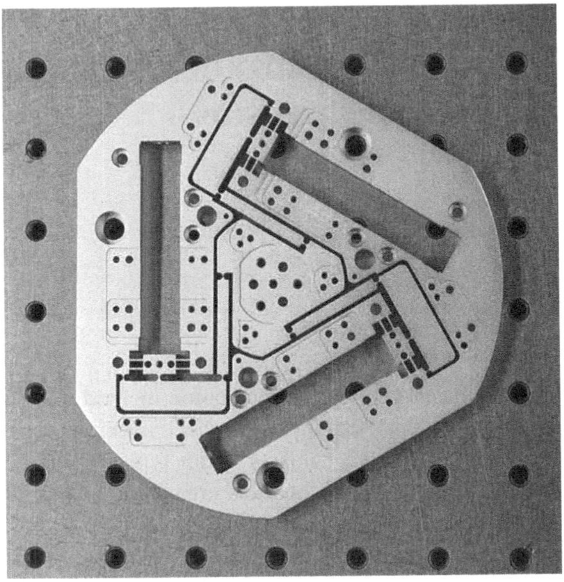

Fig. 4.1 A CPM based on a rigid-body 3-RRR mechanism [38]

rigid-body counterparts [13, 28, 36, 38, 41, 42]. To a great extent, the CPMs are interpretations of rigid-body parallel mechanisms in the field of compliant mechanisms. The conceptual synthesis of CPMs depends on the methods of synthesizing [9, 12, 21, 40] rigid-body parallel mechanisms. This dependence may lead to the following problems:

- As we know, selecting the optimal topology of a mechanism is very important because the topology essentially determines the performance of the mechanism. A poor topology may yield insufficient performance despite highly developed design tools and optimization techniques that may be used later in the development process [33]. However, the synthesis methods for rigid-body parallel mechanisms focus on enumerating all possible topologies and are currently unable to solve the problem of topology selection. Therefore, designers cannot use the best rigid-body mechanism topology to design a CPM for a particular task.
- The kinematic topologies obtained by means of rigid-body synthesis methods may not be suitable for compliant mechanisms. After all, force and displacement in a compliant mechanism are inseparable and must be considered simultaneously [4], which makes analyzing them more complex than analyzing rigid-body mechanisms.
- The CPMs obtained using the rigid-body replacement approach are limited to mechanical topologies; their topological diversity can be enriched by introducing a type of structural topology. Figure 4.2 shows the differences between mechanical and structural topologies.

Fig. 4.2 Comparison of mechanical and structural topologies

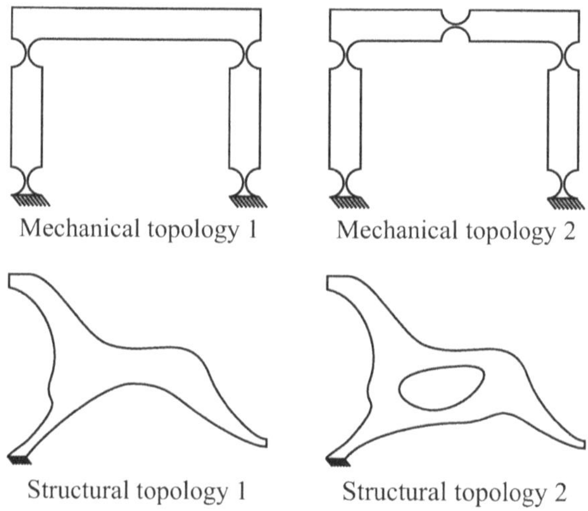

As two methods without this dependence, the freedom and constraint topology (FACT) approach proposed by Hopkins et al. [10] and the screw theory-based method proposed by Yu et al. [43] are new contributions to CPM type synthesis. However, the performance of such topologies is not considered in either method. Consequently, a systematic synthesis method that considers the topology performance is needed for CPMs.

4.2 Design of CPMs Using the Traditional Topology Optimization Method

From a structural perspective, topology optimization provides a good solution to the problem of topology selection. This approach provides the optimal topology, shape, and dimensions of a single-piece compliant mechanism directly from the design specifications, without beginning with a known rigid-body mechanism [7, 32]. In other words, this approach combines the two steps in the conceptual design of mechanisms, i.e., type synthesis and dimensional synthesis, into one step to select the best configuration for a specific design problem. Compliant mechanisms designed using this approach have structural topologies, unlike mechanisms obtained using the rigid-body replacement approach.

4.2.1 Multi-criterion Topology Optimization

Due to its advantages, the topology optimization approach has been successfully applied to the problem of synthesizing mechanisms with multiple input and output ports. For example, Frecker et al. [8] extended topology optimization to compliant mechanisms with multiple output ports using a multi-criterion formulation. Sigmund [34] designed multi-physics and multi-material actuators with multiple input and output ports using topology optimization. Saxena [31] studied topology optimization of large-displacement compliant mechanisms with multiple materials and multiple output ports. Lu and Kota [24, 25] studied topology optimization with non-fixed ports and proposed the load path representation, in which the input, output, and ground ports can have different coordinates. Zhang et al. [44] proposed a method for optimizing the topology of a compliant mechanism with multiple inputs and multiple outputs using level set methods.

Topology optimization has also been extended to the synthesis of 3D compliant mechanisms [1, 14, 39], but research has focused on the case of single input and output ports. On the whole, the multi-criterion topology optimization discussed above is for controlling the displacements of multiple output ports subject to the excitation of one or more inputs.

Figure 4.3a shows a general design domain Ω with multiple input and output requirements. f_{ai} $(i = 1, 2, \ldots, n)$ are the input forces, and u_{outj} $(j = 1, 2, \ldots, n_o)$ are the expected output displacements, where n and n_o are the numbers of inputs and outputs, respectively. For flexibility design, the mechanism's flexibility is defined in terms of the mutual strain energy (MSE). As shown in Fig. 4.3a, dummy loads f_{oj} $(j = 1, 2, \ldots, n_o)$ are applied at the output ports. The mutual strain energy E_{ms} of the compliant system is

$$E_{ms} = \sum_{i=1}^{n} \sum_{j=1}^{n_o} \mathbf{U}_{oj}^{\mathrm{T}} \mathbf{K} \mathbf{U}_{ai}, \tag{4.1}$$

where $\mathbf{U}_{oj}$ and $\mathbf{U}_i$ are the nodal displacement vectors due to the dummy load f_{oj} $(j = 1, 2, \ldots, n_o)$ and the applied load f_{ai} $(i = 1, 2, \ldots, n)$, respectively.

For stiffness design, the stiffness is defined in terms of the strain energy (SE). Figure 4.3b shows a schematic for stiffness design. The input ports are considered fixed, and loads f_{oj} $(j = 1, 2, \ldots, n_o)$ are applied at the output ports in the opposite directions, which accounts for the resistance of the working force. The strain energy of the system E_s is

$$E_s = \mathbf{U}_d^{\mathrm{T}} \mathbf{K} \mathbf{U}_d, \tag{4.2}$$

where $\mathbf{U}_d$ is the nodal displacement vector due to dummy loads $-f_{oj}$ $(j = 1, 2, \ldots, n_o)$. The objective function of the multiple input and output problem is expressed as follows:

$$\min: \quad \mu E_s - (1 - \mu) E_{ms}, \tag{4.3}$$

where $\mu \in [0, 1]$ is a weight factor.

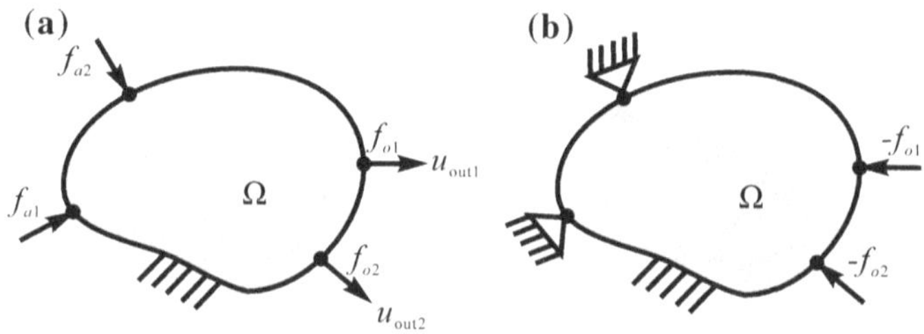

Fig. 4.3 Schematics for flexible (**a**) and stiff (**b**) designs of a mechanism with multiple input and output ports

4.2.2 Design Example

In this section, an example is presented to demonstrate the validity of the aforementioned method. A CPM with four input ports and four output ports is considered. The directions and positions of the input and output ports are shown in Fig. 4.4. The design objective is to maximize the output displacement at output i ($i = 1, 2, 3, 4$) due to a load f_i applied at input i. This is equivalent to maximizing the MSE between input i and output i.

The amount of material allowed and the weight factor μ are set to 20% and 0.3, respectively. This problem is solved using the level set method [44]. The optimized topology is shown in Fig. 4.5a. When the four outputs are in the directions shown in Fig. 4.4, the mechanism generates a rotational motion (Fig. 4.5b). As shown in Fig. 4.5c–d, translational motion is generated if the two outputs along this axis are in the same direction. One can see that the multi-criterion topology optimization method results in multiple-DOF motion using multiple specific predefined output displacements.

Fig. 4.4 Design domain for a mechanism with four inputs and four outputs

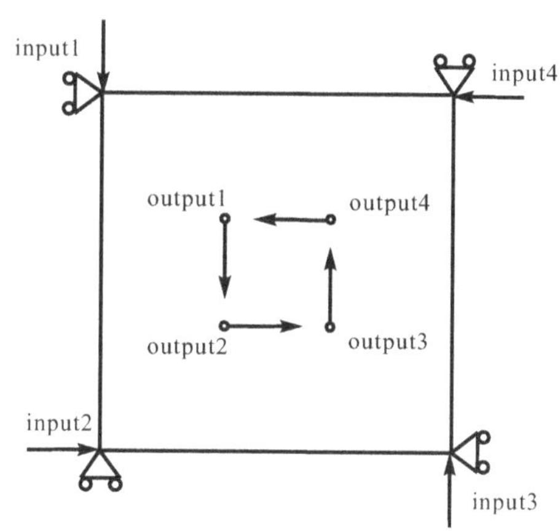

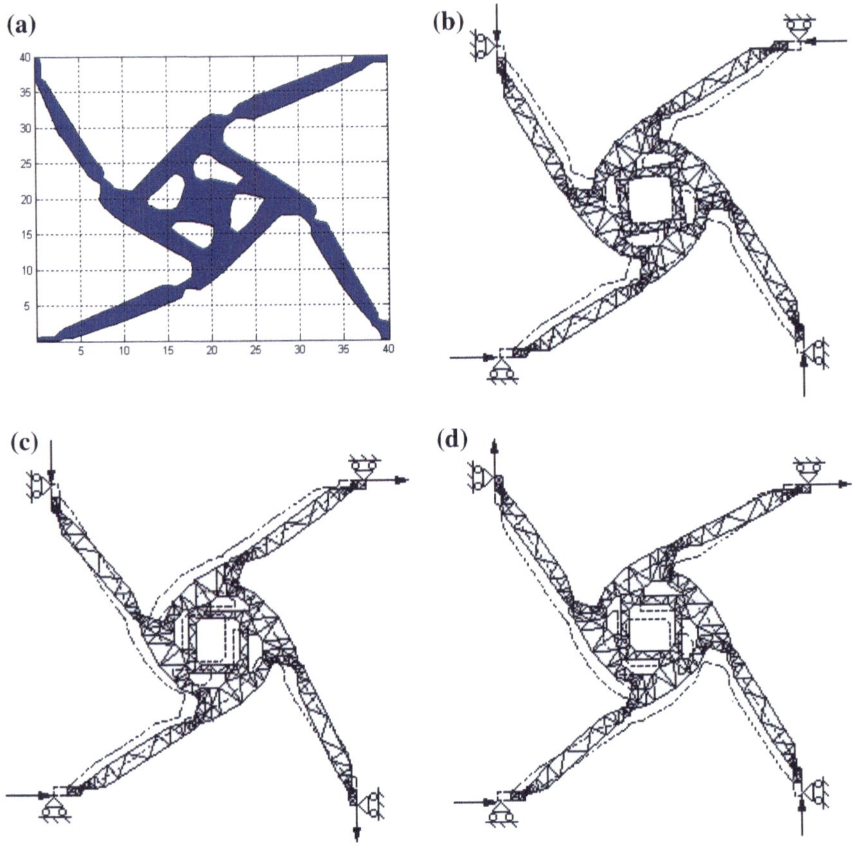

Fig. 4.5 Mechanism with four inputs and four outputs: **a** the final topology, **b** rotational output, **c** translational output along the x-axis, and **d** translational output along the y-axis

4.3 Jacobian-Based Topology Optimization

This section is to introduce the Jacobian-based topology optimization method [15]. This method is a new extension of the topology optimization techniques that employs a formulation based on the Jacobian and stiffness matrices to evaluate the overall performances of multi-DOF CPMs [15]. Unlike traditional CPMs, CPMs obtained using the proposed method have structural topologies, which is a new type of topology for CPMs.

4.3.1 Motivation

The number of DOFs at the output point of the platform is more important to the synthesis of a CPM than the specific displacements of multiple output ports. The number of DOFs of the platform can be up to three for 2D CPMs and six for 3D CPMs. It is not easy to use specific displacements to describe multi-DOF motion any more, especially in 3D cases. Therefore, topology optimization is more complex for a CPM than for a compliant mechanism with multiple input and output ports. Multi-criterion topology optimization methods cannot be used to directly synthesize CPMs.

To overcome this difficulty, our previous studies [16–18] made some attempts at the optimal synthesis of flexure-based compliant mechanisms using a new method. Reference [18] synthesized multi-DOF passive compliant mechanisms by optimizing the compliance characteristics of a mobile platform. References [16, 17] proposed a new method for synthesizing 1-DOF pseudo-rigid-body mechanisms based on the idea of topology optimization. Heading in a similar direction, Lum et al. [26, 27] recently presented a hybrid topological and structural optimization method to design 3-DOF CPMs. This method first synthesizes compliant joints with the optimal stiffness characteristics using topology optimization. The optimal compliant joints are then assembled to form a CPM using the topology of an existing rigid-body mechanism. Recently, we proposed a method for the optimal synthesis of planar CPMs in which the finite element analysis and matrix methods are combined to analyze the CPMs' Jacobian and stiffness matrices. A new formulation is proposed for topology optimization of multi-degree-of-freedom (DOF) compliant mechanisms [15].

4.3.2 Basic Idea

A parallel mechanism is a closed-loop mechanism in which a fixed base and a mobile platform are connected in parallel by multiple independent limbs. The topology of a CPM is determined by the number, topology structures, and arrangement of its compliant limbs. The compliant limbs can be arranged symmetrically or asymmetrically, and their topologies may or may not be identical. In the preliminary analysis, only the topology structures of the compliant limbs are regarded as design objects in the topology optimization problem. The other two factors, i.e., the number and arrangement of the compliant limbs, are predefined by the designer according to the problem's specifics.

A general design domain for planar CPMs is shown in Fig. 4.6. A CPM comprises n compliant limbs synthesized within multiple design domains, $\Gamma = \{\Omega_1, \Omega_2, \ldots, \Omega_n\}$. Unlike the conventional topology optimization approach, in this approach, the design domains are discretized by quadrilateral elements with three DOFs (u_x, u_y, θ_z) at each node. The popular simplified isotropic material with penalization (SIMP) scheme [2] is used to parametrize the design domains. The design variables corresponding to these design domains are illustrated by $\mathbf{X} = \{\mathbf{X}_1, \mathbf{X}_2, \ldots, \mathbf{X}_n\}$.

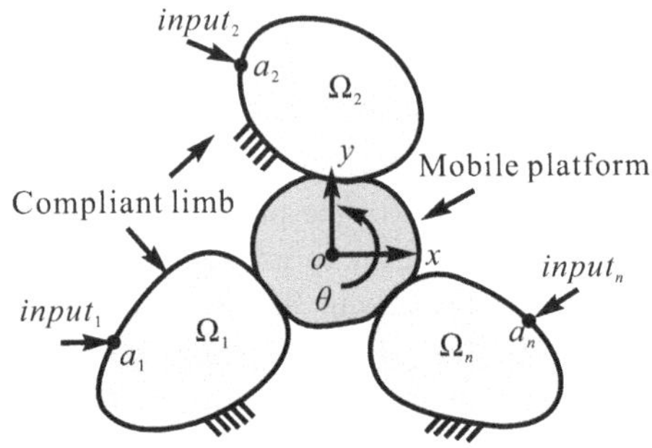

Fig. 4.6 General design domain of planar CPMs

For CPMs with identical compliant limbs, which are more common than other CPMs, the design domains and design variables are identical, e.g., $\Gamma = \{\Omega_1\}$, $\mathbf{X} = \{\mathbf{X}_1\}$.

The problem of optimizing the topology of a CPM involves finding the optimal topologies of the compliant limbs within design domain Γ based on the performance of the overall mechanism, i.e., the Jacobian and stiffness matrices of the mechanism. Kinematically analyzing CPMs with changing limb topologies is a challenging step in developing a topology optimization method for CPMs.

4.3.3 *Topology Analysis*

This section shows how the Jacobian and stiffness matrices of a CPM can be obtained using FEA and the matrix method [22, 23]. The process of modeling a CPM's stiffness is shown in Fig. 4.7. Firstly, the compliant limbs are analyzed separately to build stiffness models in their local coordinates o_i. Secondly, the stiffness matrices of the compliant limbs are transferred into global frame o. Thirdly, the stiffness matrices of the compliant limbs are combined into the mechanism stiffness matrix [37], which describes the stiffness relationships between actuated points a_i $(i = 1, 2, \ldots, n)$ and output point o. Finally, the kinematics of the CPM is calculated based on the mechanism stiffness matrix.

4.3.3.1 The Matrix Method

The matrix method is widely used in stiffness modeling and kinematic analysis of flexure hinge-based CPMs. The coordinate transformation in the matrix method is briefly reviewed here. It is well known that a 3×3 compliance matrix can be used to characterize the free end of a planar flexure in its local coordinate system. In an open chain, as shown in Fig. 4.8a, the compliance $\mathbf{C}_a$ at point a can be transferred to another frame o_1 by a transformation matrix $\mathbf{T}_a^{o_1}$;

$$\mathbf{C}_a^{o_1} = \mathbf{T}_a^{o_1} \otimes \mathbf{C}_a, \tag{4.4}$$

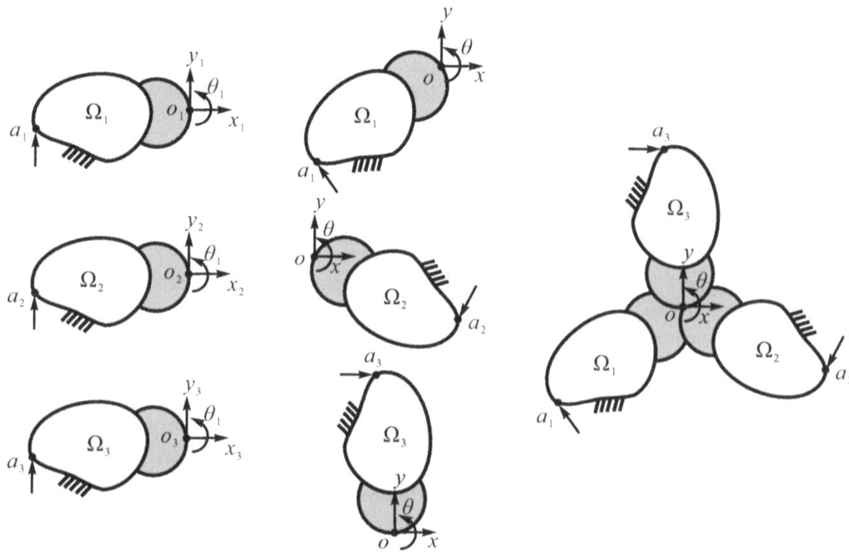

Fig. 4.7 Stiffness modeling process for CPMs

where $\otimes$ is defined as

$$\mathbf{A} \otimes \mathbf{C} = \mathbf{A}\mathbf{C}\mathbf{A}^{\mathrm{T}}. \tag{4.5}$$

The transformation matrix $\mathbf{T}_a^{o_1}$ is

$$\mathbf{T}_a^{o_1} = \mathbf{P}(\mathbf{r}_a^{o_1})\,\mathbf{R}(\theta_a^{o_1}), \tag{4.6}$$

where $\mathbf{P}(\mathbf{r}_a^{o_1})$ is the translational transformation matrix expressed in Eq. (4.7), $\mathbf{r}_a^{o_1}$ ($\mathbf{r} = [r_x,\ r_y]$) is the position vector of point a in frame o_1, $\mathbf{R}(\theta_a^{o_1})$ is the rotational transformation matrix, and $\theta_a^{o_1}$ is the angle of frame a with respect to frame o_1.

$$\mathbf{P}(\mathbf{r}) = \begin{bmatrix} 1 & 0 & r_y \\ 0 & 1 & -r_x \\ 0 & 0 & 1 \end{bmatrix}. \tag{4.7}$$

For the serially connected flexures in an open chain, the compliance of each endpoint is the superposition of all the flexures' compliances at one point. In Fig. 4.8a, the compliance of endpoint o_1 is the superposition of the compliances of flexure hinges a and b at the coordinates of point o_1, which can be expressed as

$$\mathbf{C}_{o_1} = \mathbf{T}_a^{o_1} \otimes \mathbf{C}_a + \mathbf{T}_b^{o_1} \otimes \mathbf{C}_b. \tag{4.8}$$

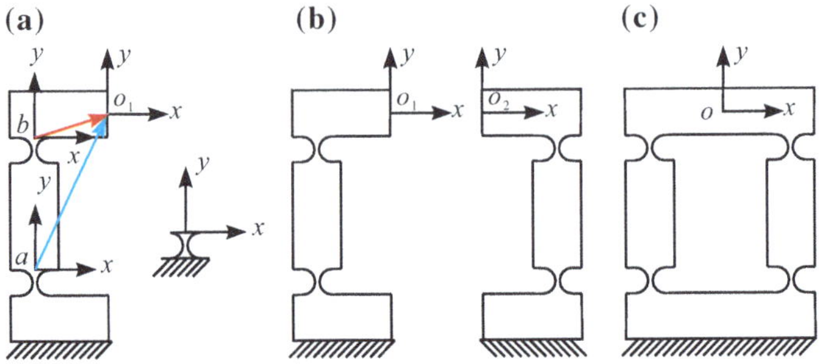

Fig. 4.8 Coordinate transformation using the matrix method

For the chains connected in parallel, the stiffness of each junction is the superposition of all the chains' stiffnesses at one point. The compliant 4-bar mechanism shown in Fig. 4.8c is composed of the two chains in Fig. 4.8b. Therefore, the stiffness at point o is the superposition of the stiffnesses of endpoints o_1 and o_2 at point o, i.e.,

$$\mathbf{K}_o = \mathbf{T}^o_{o_1} \otimes \mathbf{K}_{o_1} + \mathbf{T}^o_{o_2} \otimes \mathbf{K}_{o_2}. \tag{4.9}$$

4.3.3.2 Modeling the Stiffness of CPMs

One advantage of using multiple design domains is that the compliant limbs of a CPM can be discretized, parametrized and analyzed separately in their local coordinate systems. A compliant limb and its local coordinate system is shown in Fig. 4.9; it consists of design domain Ω_i and part of mobile platform Ω^i_p. The global stiffness matrix of the ith compliant limb can be obtained using Eq. (4.10),

$$\begin{aligned} &\mathbf{K}_i(\mathbf{X}_i) = \sum_{e=1}^{N_i} (x^e)^\rho \mathbf{K}_e + \sum_{e=1}^{N^i_p} \mathbf{K}_e \\ &0 < x^e_{min} \leq x^e \leq 1, \quad x^e \in \mathbf{X}_i, \end{aligned} \tag{4.10}$$

where N_i is the number of elements in design domain Ω_i, N^i_p is the number of elements in Ω^i_p, $\mathbf{K}_e$ is the global element stiffness matrix, x^e is the material density (design variable) of each element in Ω_i with whose value is between the lower limit x^e_{min} (void) and 1 (solid), ρ is the penalty factor, and the elements in Ω^i_p are solid.

For a compliant limb in its local coordinate system (as shown in Fig. 4.9), only the number of input DOFs at point a_i and the three DOFs of endpoint o_i are considered in stiffness modeling. First, a compliance matrix $\mathbf{C}^i_{ao}$ that characterizes the compliance relationship among these four concerned DOFs is calculated based on the physical meaning of compliance. Four different loads $\mathbf{F}_j$ ($j = 1, 2, \ldots, 4$), in which a unit

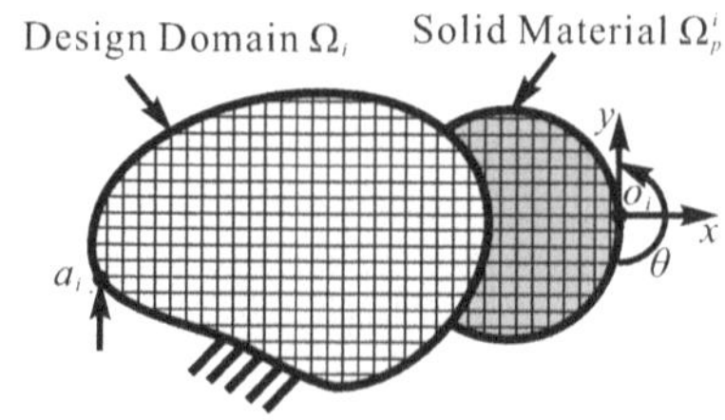

Fig. 4.9 Stiffness modeling schematic for the ith compliant limb

dummy load is applied for each relevant DOF of a_i and o_i in sequence, are used to calculate corresponding displacements by solving Eq. (4.11),

$$\mathbf{F}_j = \mathbf{K}_i \mathbf{U}_j, \quad j = 1, 2, \ldots, 4. \tag{4.11}$$

The physical meaning of element C_{jk} in a compliance matrix $\mathbf{C}$ is the displacement of the jth DOF due to a unit load that acts only on the kth DOF. Based on this physical meaning, the element of $\mathbf{C}_{ao}^i$ is obtained using Eq. (4.12). The nodal displacements of the four relevant DOFs in displacement vector $\mathbf{U}_k$ form the kth column of $\mathbf{C}_{ao}^i$,

$$\mathbf{C}_{ao}^i(j, k) = \mathbf{F}_j^{\mathrm{T}} \mathbf{U}_k, \quad j, k = 1, 2, \ldots, 4. \tag{4.12}$$

Then, the compliance of endpoint o_i is transferred to the coordinates of output point o by transformation matrix $\mathbf{T}_i^o$, whereas the compliance of the input DOF at a_i remains in its local coordinates. The inverse of the resulting compliance matrix, i.e., stiffness matrix $\mathbf{K}_{ao}^i$, is shown in Eq. (4.13),

$$\mathbf{K}_{ao}^i = \left(\begin{bmatrix} 1 & \mathbf{0} \\ \mathbf{0} & \mathbf{T}_i^o \end{bmatrix} \otimes \mathbf{C}_{ao}^i \right)^{-1}, \tag{4.13}$$

where 1 means that the compliance of the input DOF at a_i remains in its local coordinates.

Finally, the stiffness model of the compliant limbs is incorporated into the stiffness model of the CPM. The stiffnesses at output point o in $\mathbf{K}_{ao}^i$ for all the compliant limbs are superimposed to determine the stiffness at output point o on the mobile platform, while the stiffnesses of all the input DOFs at points a_i ($i = 1, 2, \ldots, n$) remain in their local coordinates. The transformation is shown in Eq. (4.14). The resulting $\mathbf{K}_m$ is the mechanism stiffness matrix, which characterizes the stiffness relationship between the n input DOFs at a_i and output point o,

$$\mathbf{K}_m = \begin{bmatrix} 1\,0\,0\,0 \cdots 0\,0 \\ 0\,0\,1\,0 \cdots 0\,0 \\ \vdots\;\vdots\;\vdots\;\vdots\;\ddots\;\vdots\;\vdots \\ 0\,0\,0\,0 \cdots 1\,0 \\ \mathbf{0}\,\mathbf{I}\,\mathbf{0}\,\mathbf{I} \cdots \mathbf{0}\,\mathbf{I} \end{bmatrix} \otimes \begin{bmatrix} \mathbf{K}_{ao}^1 & \mathbf{0} & \cdots & \mathbf{0} & \mathbf{0} \\ \mathbf{0} & \mathbf{K}_{ao}^2 & \cdots & \mathbf{0} & \mathbf{0} \\ \vdots & \vdots & \ddots & \vdots & \vdots \\ \mathbf{0} & \mathbf{0} & \cdots & \mathbf{K}_{ao}^{n-1} & \mathbf{0} \\ \mathbf{0} & \mathbf{0} & \cdots & \mathbf{0} & \mathbf{K}_{ao}^n \end{bmatrix}, \tag{4.14}$$

where 1 means that the compliance of the n input DOFs at a_i remain in their local coordinates and $\mathbf{I}$ is a 3×3 identity matrix. The last row of this transformation matrix combines the stiffnesses at the ends of all the compliant limbs into a single stiffness. In $\mathbf{K}_m$, the diagonal elements define the characteristic stiffness of the compliant mechanism [5, 37]. The elements related to the input DOFs are the input characteristic stiffness of the CPM and denoted by k_{in}^{i} $(i = 1, 2, \ldots, n)$. In contrast, the diagonal elements related to the output DOFs are the output characteristic stiffness of the CPM and denoted by k_{out}^{j} $(j = 1, 2, 3)$.

The characteristic stiffness is considered an intrinsic property of the compliant system. It is independent of the external load and evaluates the essential stiffness at a specific port, which makes it different from the frequently-used strain energy. By maximizing the input and output characteristic stiffnesses, the stiffness requirements of compliant mechanisms can be satisfied during topology optimization.

4.3.3.3 Kinematic Analysis of CPMs

Based on the input and output DOFs of the CPM, the mechanism's displacements can be partitioned into sets $\mathbf{U}_a$ and $\mathbf{U}_o$ for the displacements of the input and output DOFs, respectively. As shown in Eq. (4.15), the loads on the mechanism are also partitioned into two sets, $\mathbf{F}_a$ and $\mathbf{F}_o$. This effectively partitions the mechanism's stiffness matrix $\mathbf{K}_m$ as follows:

$$\begin{bmatrix} \mathbf{F}_a \\ \mathbf{F}_o \end{bmatrix} = \underbrace{\begin{bmatrix} \mathbf{K}_{11} & \mathbf{K}_{12} \\ \mathbf{K}_{21} & \mathbf{K}_{22} \end{bmatrix}}_{\mathbf{K}_m} \begin{bmatrix} \mathbf{U}_a \\ \mathbf{U}_o \end{bmatrix}. \tag{4.15}$$

Assuming that no external load is applied at the output point o of the mobile platform, i.e., $\mathbf{F}_o = \mathbf{0}$,

$$\mathbf{F}_o = \mathbf{K}_{21}\mathbf{U}_a + \mathbf{K}_{22}\mathbf{U}_o = \mathbf{0}. \tag{4.16}$$

By solving the above equation, the relationship between the input displacement $\mathbf{U}_a$ and the output displacement $\mathbf{U}_o$, i.e., the Jacobian matrix $\mathbf{J}$, can be obtained;

$$\mathbf{J} = -\mathbf{K}_{22}^{-1}\mathbf{K}_{21}. \tag{4.17}$$

Suppose $\mathbf{U}_o = [u_x, u_y, \theta_z]^{\mathrm{T}}$ and $\mathbf{U}_a = [u_1, u_1, \ldots, u_n]^{\mathrm{T}}$. The specific form of the Jacobian matrix is as follows:

$$\begin{bmatrix} u_x \\ u_y \\ \theta_z \end{bmatrix} = \underbrace{\begin{bmatrix} J_{11} & J_{12} & \cdots & J_{1n} \\ J_{21} & J_{22} & \cdots & J_{2n} \\ J_{31} & J_{32} & \cdots & J_{3n} \end{bmatrix}}_{\mathbf{J}} \begin{bmatrix} u_1 \\ u_2 \\ \vdots \\ u_n \end{bmatrix}. \tag{4.18}$$

With Jacobian matrix $\mathbf{J}$, the workspace of the CPM can be analyzed based on a limited input displacement $u_i \in [0, \delta]$, where δ is the stroke length of the actuator. With the workspace along the x-axis as an example, the maximum and minimum output displacements in the x-direction are presented in Eq. (4.19). Suppose that the jth ($j = 1, 2, 3$) DOF of $\mathbf{U}_o$ is maximized when the input is $\mathbf{U}'_a$ and minimized when the input is $\mathbf{U}''_a$,

$$\begin{aligned} U_{oj}^{max} &= J_{11}u'_1 + J_{12}u'_2 + \cdots + J_{1n}u'_n \\ U_{oj}^{min} &= J_{11}u''_1 + J_{12}u''_2 + \cdots + J_{1n}u''_n. \end{aligned} \tag{4.19}$$

Since U_{oj} is a linear function of the input displacements $\mathbf{U}_a$, u'_i and u''_i in Eq. (4.19) have values of 0 or δ according to the sign of J_{1i}. As shown in Eq. (4.20), the workspace of the jth ($j = 1, 2, 3$) DOF is determined by the sum of the absolute values of the corresponding elements in $\mathbf{J}$,

$$\Delta U_{oj} = U_{oj}^{max} - U_{oj}^{min} = \sum_{i=1}^{n} J_{ji}(u'_i - u''_i) = (\sum_{i=1}^{n} |J_{ji}|)\,\delta. \tag{4.20}$$

Element J_{ji} of $\mathbf{J}$ represents the geometric advantage (GA) between the jth DOF of $\mathbf{U}_o$ and the ith input. If all the elements of the jth row in $\mathbf{J}$ are equal or close to zero, no input $\mathbf{U}_a$ produces a displacement in the jth DOF of the mobile platform, i.e., this DOF is suppressed. In contrast, the DOFs that correspond to non-zero row vectors are considered the CPM's DOFs. By maximizing the absolute value of the elements in the non-zero row vectors, the forward kinematics and motion transmission performance of the CPM can be optimized. Therefore, the Jacobian matrix contains information about the CPM's function and performance simultaneously. By optimizing the Jacobian matrix, we can synthesize a compliant mechanism with the desired DOFs to that is functional and optimize its kinematics directly to improve its performance.

4.3.4 Optimization Model

4.3.4.1 Problem Formulation

Optimizing the topology of a compliant mechanism is actually a multi-objective optimization problem in which the designer must consider compliance and stiffness simultaneously to ensure that the mechanism is flexible enough to move and stiff enough to bear the external load. The objective function formulation is inspired by the work of Chen et al. [5], in which the objective function comprises two maximization objectives, i.e., the function constraint f and the strength constraint S. The function constraint f expresses the kinematic requirements of the CPM based on the Jacobian matrix $\mathbf{J}$, while the strength constraint S requires maximization of the characteristic input and output stiffnesses. Usually, multi-objective optimization problems are converted to single-objective optimization problems. The general problem formulation adopted is as follows:

$$\begin{aligned} \text{min}: \quad & \zeta(\mathbf{X}) = -f^{\omega} S^{(1-\omega)} \\ \text{s.t.}: \quad & V(\mathbf{X}) \leq V_o, \end{aligned} \tag{4.21}$$

where ω $(0 < \omega < 1)$ is a weight indicating the relative significance of the kinematic constraint f, $V(\mathbf{X})$ is the volume fraction of a candidate topology, and V_o is the allowed volume fraction.

The kinematic constraint f has two different forms according to the design problem. The first form of kinematic constraint, f_1, is suitable for CPMs with kinematics that are simple enough to be predefined by designers. In this form, the Jacobian matrix $\mathbf{J}$ of the CPM is forced to be close to a desired Jacobian matrix $\mathbf{J}^*$ by minimizing the difference between the elements of $\mathbf{J}$ and $\mathbf{J}^*$. As a result, the desired DOFs and kinematic properties of the CPM can be expressed by the desired Jacobian matrix $\mathbf{J}^*$,

$$\begin{aligned} \text{max}: \quad & f_1 = e^{-\sum (J_{ji} - J_{ji}^*)^2} \\ & j = 1, 2, \ldots, 3 \,\&\, i = 1, 2, \ldots, n. \end{aligned} \tag{4.22}$$

The second form of kinematic constraint, f_2, is suitable for CPMs with complex kinematics. This form is for maximizing the workspaces of the desired DOFs and suppressing the other DOFs;

$$\text{max}: \quad f_2 = e^{-\sum J_{ji}^2} \prod_{j=1}^{n} J_j, \tag{4.23}$$

where J_j evaluates the workspace in the jth desired DOF and J_j is the sum of the squares of the elements in the corresponding row vector,

$$J_j = \sum_{i=1}^{n} J_{ji}^2. \tag{4.24}$$

The stiffness constraint S requires maximization of the characteristic input and output stiffnesses of the CPM. Mathematically, the stiffness constraint is formulated as the product of the characteristic input stiffness k_{in}^{i} and the characteristic output stiffness k_{out}^{i},

$$\max: \quad S = \prod_{i=1}^{n} k_{in}^{i} k_{out}^{i}. \tag{4.25}$$

4.3.4.2 Sensitivity Analysis

Solving the optimization problem requires determination of the sensitivities. The sensitivity of each objective function discussed above is determined by the sensitivities of the Jacobian matrix $\mathbf{J}$ and mechanism stiffness matrix $\mathbf{K}_m$ because J_{ji}, k_{in}^{i}, and k_{out}^{j} are the elements of $\mathbf{J}$ and $\mathbf{K}_m$, respectively. According to Eq. (4.17), the sensitivity of $\mathbf{J}$ is

$$\frac{\partial \mathbf{J}}{\partial x^e} = \mathbf{K}_{22}^{-1} \frac{\partial \mathbf{K}_{22}}{\partial x^e} \mathbf{K}_{22}^{-1} \mathbf{K}_{21} - \mathbf{K}_{22}^{-1} \frac{\partial \mathbf{K}_{21}}{\partial x^e}. \tag{4.26}$$

$\mathbf{K}_{22}$ and $\mathbf{K}_{21}$ are parts of stiffness matrix $\mathbf{K}_m$; therefore, their sensitivities can be extracted from the sensitivity of $\mathbf{K}_m$. According to Eq. (4.14), the sensitivity of $\mathbf{K}_m$ is determined by the sensitivities of $\mathbf{K}_{ao}^{i}$,

$$\frac{\partial \mathbf{K}_{ao}^{i}}{\partial x^e} = -\mathbf{K}_{ao}^{i} \left(\begin{bmatrix} 1 & \mathbf{0} \\ \mathbf{0} & \mathbf{T}_i^o \end{bmatrix} \otimes \frac{\partial \mathbf{C}_{ao}^{i}}{\partial x^e} \right) \mathbf{K}_{ao}^{i}. \tag{4.27}$$

The sensitivity of $\mathbf{C}_{ao}^{i}$ is

$$\begin{aligned} \frac{\partial \mathbf{C}_{ao}^{i}(j,k)}{\partial x^e} &= \mathbf{F}_j^{\mathrm{T}} \frac{\partial \mathbf{U}_k}{\partial x^e} = \mathbf{F}_j^{\mathrm{T}} \frac{\partial}{\partial x^e} (\mathbf{K}_i^{-1} \mathbf{F}_k) \\ &= -\mathbf{F}_j^{\mathrm{T}} \mathbf{K}_i^{-1} \frac{\partial \mathbf{K}_i}{\partial x^e} \mathbf{K}_i^{-1} \mathbf{F}_k \\ &= -\mathbf{U}_j^{\mathrm{T}} \frac{\partial \mathbf{K}_i}{\partial x^e} \mathbf{U}_k \\ &= -p (x^e)^{p-1} \mathbf{U}_j^{\mathrm{T}} \mathbf{K}_e \mathbf{U}_k. \end{aligned} \tag{4.28}$$

4.3.4.3 Optimization Algorithm

A flowchart for the optimization algorithm is shown in Fig. 4.10. First, the design domains are initialized to finite elements with a given material volume. After initialization, the stiffness matrix of each compliant limb is obtained to characterize points a_i and o in their local coordinates. Then, the stiffness of each compliant limb is transformed to the global coordinate system and all are assembled to build the stiffness model of the CPM using the matrix method. The objective function is formulated

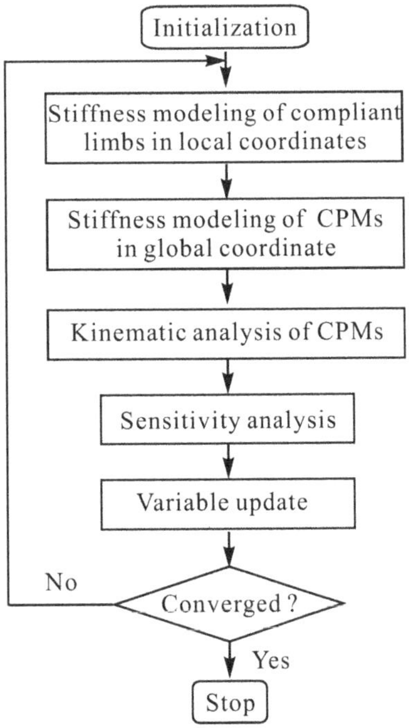

Fig. 4.10 Flowchart of the optimization algorithm

using the Jacobian and stiffness matrices of the CPM. Following this, a sensitivity analysis is conducted and the variables are updated using the OC-method. When the convergence condition is satisfied, the resulting topology is graphed.

4.3.5 Case Studies

In this section, two sample CPM designs are used to illustrate the Jacobian-based topology optimization method. Additionally, the method is applied to the benchmark problem for traditional topology optimization methods. The artificial material properties for these examples are as follows: the Young's modulus is $E = 1$ GPa, and the Poisson's ratio is $\upsilon = 0.3$.

4.3.5.1 2-DOF CPM

As shown in Fig. 4.11, the first problem is to synthesize a 2-DOF CPM with a pair of asymmetrical compliant limbs that have the same size and boundary conditions. The two compliant limbs, $limb1$ and $limb2$, are considered the design domains of the topology optimization problem. The output point o is at the center of the mobile

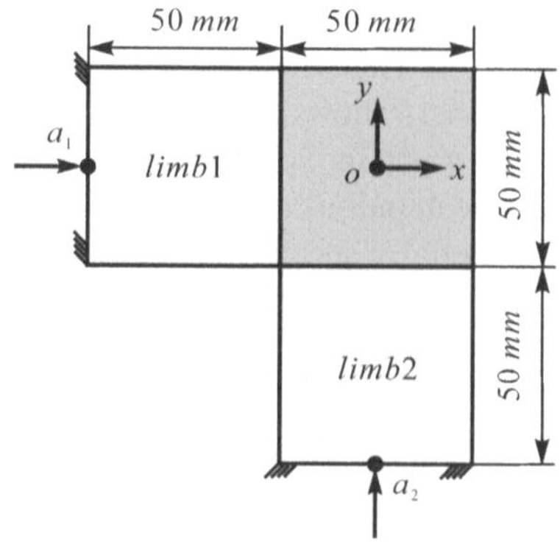

Fig. 4.11 Design domain for a 2-DOF CPM

platform, and points a_1 and a_2 are the input points of the CPM. The design domain is discretized into 100×100 finite elements for elastic analysis. The allowable amount of material is 20%.

Since the kinematics of 2-DOF CPMs is relatively simple compared with that of 3-DOF CPMs, it is easy to synthesize a 2-DOF CPM with the desired $\mathbf{J}^*$ using the first form of the kinematic constraint, f_1. The $\mathbf{J}^*$ for this example is shown in Eq. (4.29), where a_{1x} and a_{2y} are the displacements of the input DOFs. Because the third row of $\mathbf{J}^*$ is a zero vector, the DOF θ_z is suppressed; i.e., the CPM is expected to have only two translational DOFs (u_x and u_y). Moreover, these two translational DOFs are expected to be kinematically decoupled; i.e., one input only induces one translational displacement with desired geometric advantage -3 and has no impact on the other two DOFs;

$$\begin{bmatrix} u_x \\ u_y \\ \theta_z \end{bmatrix} = \underbrace{\begin{bmatrix} -3 & 0 \\ 0 & -3 \\ 0 & 0 \end{bmatrix}}_{\mathbf{J}^*} \begin{bmatrix} a_{1x} \\ a_{2y} \end{bmatrix}. \tag{4.29}$$

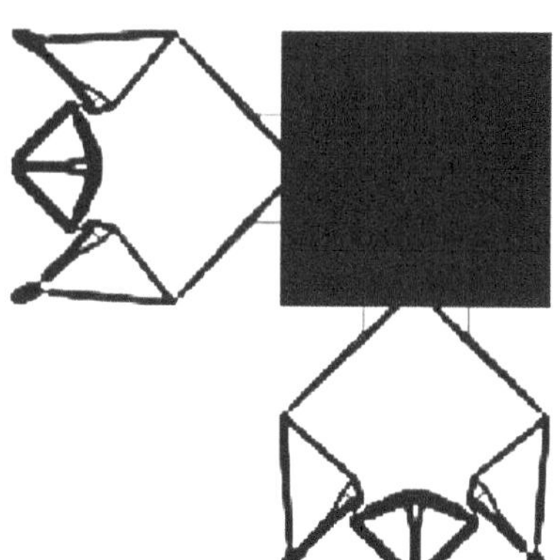

Fig. 4.12 The final topology of the 2-DOF CPM

The weight of the kinematic constraint f is set to $\omega = 0.5$. The optimization is run for 300 iterations. The final topology of the 2-DOF CPM is shown in (Fig. 4.12). Figure 4.13 shows the iteration history of the objective function, the kinematic constraint f and the stiffness constraint S during the optimization process. It can be seen that the iteration curves oscillate. The oscillations may be caused by the distribution of material at specific elements. Fortunately, the topology becomes stable later in the iteration process; this can be regarded as the optimal topology. The Jacobian matrix $\mathbf{J}$ of the final topology is shown in Eq. (4.30). One can see that the resulting Jacobian matrix is close to the desired Jacobian matrix $\mathbf{J}^*$. The elements that are expected to be equal to zero (e.g., J_{12}, J_{21}, J_{31}, and J_{32}) are close enough to zero. As a result, the rotational DOF θ_z of the CPM is suppressed. The non-zero elements indicate that the CPM has two decoupled translational DOFs, i.e., u_x and u_y. Each translational DOF has an optimized geometric advantage of -2.64. The results discussed above show that the 2-DOF CPM's function and kinematic performance can be optimized simultaneously by optimizing its Jacobian matrix. In addition, maximizing the characteristic input and output stiffnesses guarantees a valid material connection during topology optimization.

$$\mathbf{J} = \begin{bmatrix} -2.64 & -0.03 \\ 0.04 & -2.64 \\ 0.04 & -0.03 \end{bmatrix}. \tag{4.30}$$

4.3.5.2 Three-DOF CPM

The second example is the synthesis of an optimal 3-DOF CPM with three compliant limbs. As shown in Fig. 4.14, three design domains $\Gamma = \{\Omega_1, \Omega_2, \Omega_3\}$ of the CPM are symmetrically arranged and have the same size. Each design domain is discretized into 50×80 finite elements for elastic analysis with the same boundary conditions. The amount of material allowed is 30%. It is difficult to predefine a target $\mathbf{J}^*$ for a 3-DOF CPM; therefore, this type of design problem involves the second form of the kinematic constraint, f_2.

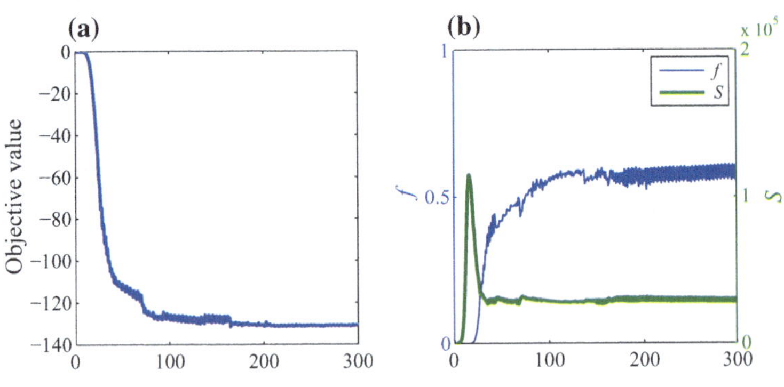

Fig. 4.13 Iteration history of the 2-DOF CPM: **a** objective function and **b** f and S

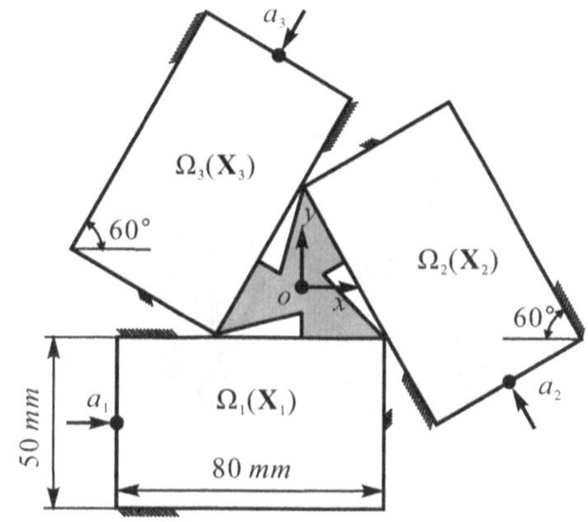

Fig. 4.14 Design domain for a 3-DOF CPM

The optimization process was run for 500 iterations. The resulting optimal topology is described in Fig. 4.15a. One can see that the three resulting compliant limbs have nearly identical topologies and that each compliant limb has only one fixed port connected by material. The iteration history of the objective function $\zeta(\mathbf{X})$, the kinematic constraint f, and the stiffness constraint S throughout the optimization process are shown in Fig. 4.16.

The theoretical Jacobian matrix $\mathbf{J}_{theo}$ for the optimal topology is shown in Table 4.1. To verify $\mathbf{J}_{theo}$, FEA in ANSYS is used to evaluate the simulated Jacobian matrix $\mathbf{J}_{sim}$, which is also shown in Table 4.1. According to the deviation between the theoretical and numerical results, $\mathbf{D}_{sim}^{theo}$, $\mathbf{J}_{theo}$ is close to $\mathbf{J}_{sim}$ except at element J_{11}, whose deviation is up to 31%. One reason for this is that the computational accuracy of the adopted element stiffness matrix is not sufficient because quadrilateral elements with one rotational DOF per node are not commonly used and require further investigation. Another reason lies in our kinematic modeling method. The compliance matrix of each compliant limb obtained from the FEA deviates slightly from the exact matrix; this deviation is caused in part by modeling precision of the matrix method.

To verify the mobility of the optimal topology, the CPM is required to output pure translational and rotational motions. A pure output motion of the mobile platform is predefined to calculate the corresponding input $\mathbf{U}_{in}$ using the inverse of $\mathbf{J}_{sim}$, i.e., the inverse kinematics of the CPM. The resulting simulations for the calculated $\mathbf{U}_{in}$ are presented in Fig. 4.15b–d. The accurate deformations of the mobile platform in these simulations show that $\mathbf{J}_{sim}$ can be used to precisely evaluate the kinematics of the CPM.

Based on the optimal topology, the prototype shown in Fig. 4.17 is manufactured using 3D printing with resin. A parallel guiding mechanism is added at each input point to guide the input motion, which avoids subjecting the piezoelectric (PZT) actuators to lateral forces. As shown in Fig. 4.17b, three PZT actuators are adopted

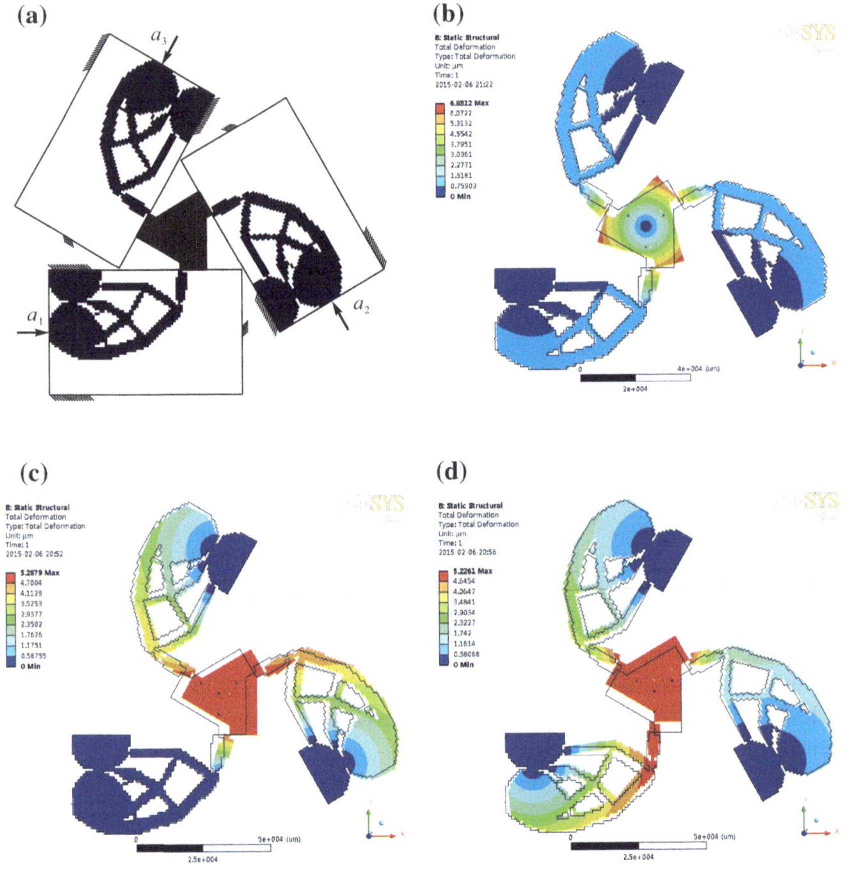

Fig. 4.15 3-DOF CPM with three compliant limbs: **a** the optimal topology, **b** rotational motion for $\mathbf{U}_{in} = [1, 1, 1]^T$, **c** translational motion along the x-axis for $\mathbf{U}_{in} = [0.21, -1.69, 1.47]^T$, and **d** translational motion along the y-axis for $\mathbf{U}_{in} = [1.82, -0.73, 1.09]^T$

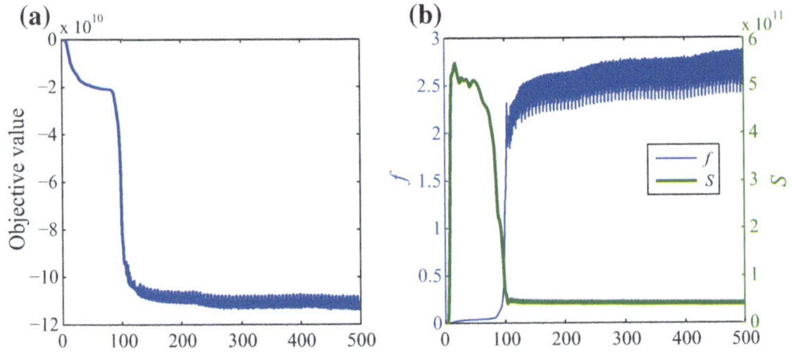

Fig. 4.16 Iteration history of the 3-DOF CPM: **a** the objective function and **b** f and S

Table 4.1 Theoretical, simulation, and experimental Jacobian matrices of the 3-DOF CPM

$\mathbf{J}$	J_{11}	J_{12}	J_{13}	J_{21}	J_{22}	J_{23}	J_{31}	J_{32}	J_{33}
$\mathbf{J}_{theo}$	0.15	−1.63	1.48	1.79	-0.77	−1.02	0.14	0.14	0.14
$\mathbf{J}_{sim}$	0.21	−1.66	1.46	1.81	-0.72	−1.09	0.13	0.13	0.13
$\mathbf{J}_{exp}$	0.24	−1.64	1.28	1.75	-0.64	−1.07	0.12	0.12	0.12
$\mathbf{D}_{sim}^{theo}$ (%)	30.5	1.6	1.4	0.7	7.4	5.9	2.3	3.2	2.5
$\mathbf{D}_{sim}^{exp}$ (%)	11.9	0.7	12.2	3.2	10.4	1.3	8.7	8.2	8.5
$\mathbf{D}_{exp}^{theo}$ (%)	37.9	0.9	15.4	2.6	19.9	4.6	11.9	12.4	12

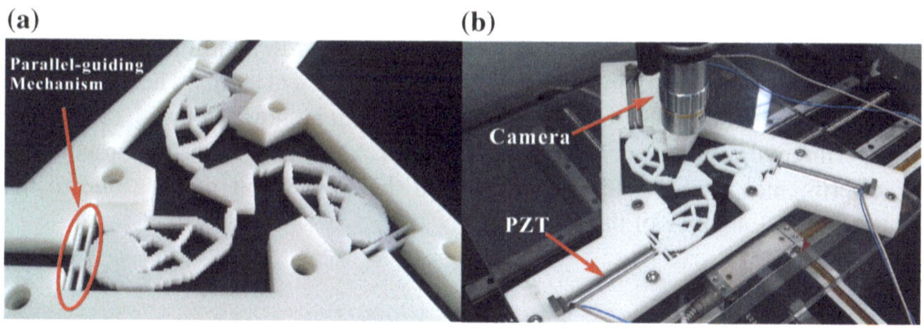

Fig. 4.17 Prototype (**a**) and experimental setup (**b**) for the 3-DOF CPM

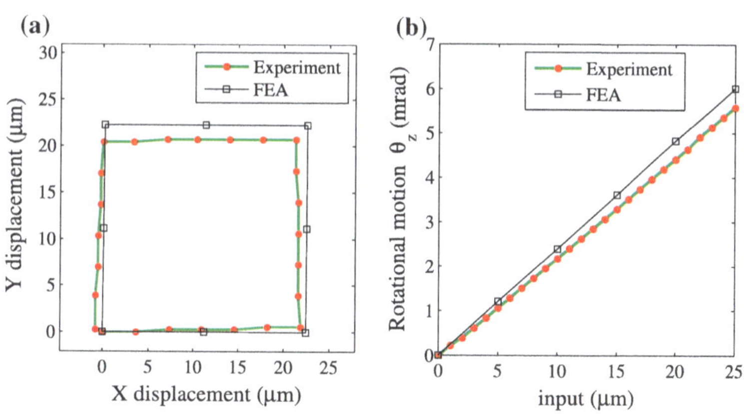

Fig. 4.18 Experimental results: translational trajectory (**a**) and rotational motion (**b**)

to drive the prototype. The output displacement at point o is measured by a micro-vision measuring system. The experiments are conducted under open-loop control.

First, the experimental Jacobian matrix $\mathbf{J}_{exp}$ of the prototype is tested. When the CPM is driven by the ith PZT actuator, the displacements of two marker points on the x- and y-axes of the mobile platform are measured by the micro-vision measuring system. With the two marker points' displacements, the translational displacements along the x- and y-axes and the rotational displacement around the z-axis of the mobile platform can be calculated. These output displacements are divided by the input displacement of the ith PZT actuator to obtain the elements of the ith column of $\mathbf{J}_{exp}$. The $\mathbf{J}_{exp}$ obtained from this approach is shown in Table 4.1. It is clear that the maximum deviation of $\mathbf{J}_{exp}$ from $\mathbf{J}_{sim}$ is approximately 12%. In contrast, the deviation of J_{11} in $\mathbf{J}_{exp}$ and $\mathbf{J}_{theo}$ is up to 38% for the reasons discussed in Sect. 4.3.5.2.

Second, the mobility of the CPM is tested. The mobile platform must translate along a rectangular trajectory. The inputs of the three PZT actuators are calculated using inverse kinematics based on $\mathbf{J}_{exp}$. The displacement of point o on the mobile platform is measured. Figure 4.18a plots the actual trajectory of point o. One can see that the actual trajectory is close to the trajectory obtained in the FEA. The deviation of the actual trajectories from those resulting from FEA may arise from the accuracy of $\mathbf{J}_{exp}$ and hysteresis of the PZT actuators. In addition, the rotational mobility is measured by driving the CPM with three identical inputs. The relationship between the pure rotational displacement and the input is shown in Fig. 4.18b, where the experimental results are compared with the FEA results. The deviation between the experimental rotational displacement and the FEA result may be due to the manufacturing tolerance of the prototype.

4.4 An Improved Stiffness Evaluation Method for Jacobian-Based Topology Optimization

In the previous section, a Jacobian-based topology optimization method for CPMs is proposed; in it, a CPM's Jacobian matrix and characteristic stiffness are optimized simultaneously to satisfy the kinematic and stiffness constraints. However, the characteristic stiffness fails to ensure that a valid topology results in certain cases. To solve this problem, an improved stiffness evaluation method based on the definition of stiffness is developed in this section. This new stiffness evaluation method is verified and the result is compared with the characteristic stiffness using several design examples. In addition, several typical benchmark problems (a displacement inverter, a displacement amplifier, and a displacement redirector) are solved using Jacobian-based topology optimization to show its general applicability.

4.4.1 Properties of the Jacobian Matrix

With consideration of the particular application, the Jacobian matrix, which has previously been used in the analysis of rigid-body mechanisms and flexure-based CPMs, is introduced to topology optimization as an alternative kinematic formulation. As shown in Eq. (4.31), the Jacobian matrix $\mathbf{J}$ describes the relationship between input and output displacements, i.e., the forward kinematics of the CPM;

$$\mathbf{U}_o = \mathbf{J}\mathbf{U}_a, \tag{4.31}$$

where $\mathbf{U}_o = [u_x, u_y, \theta_z]^{\mathrm{T}}$ contains the displacements of output point o on the mobile platform in the three directions, and $\mathbf{U}_a = [u_1, u_2, \ldots, u_n]^{\mathrm{T}}$ is the vector of input displacements at points a_i. Because compliant mechanisms are usually driven by linear actuators, we limit the displacements in $\mathbf{U}_a$ to translational motions of the input points.

Element J_{ji} of $\mathbf{J}$ represents the geometric advantage (GA) between the jth freedom of $\mathbf{U}_o$ and the ith input. If all the elements of the jth row of $\mathbf{J}$ are equal or close to zero, none of the inputs of $\mathbf{U}_a$ produce displacements of the mobile platform in the jth direction, i.e., this degree of freedom is suppressed. In contrast, the DOFs corresponding to non-zero row vectors are considered the CPM's DOFs. By maximizing the absolute values of the elements in the non-zero row vectors, the forward kinematics and motion transmission performance of the CPM can be optimized. Therefore, the Jacobian matrix contains information on the CPM's function and performance simultaneously. By optimizing the Jacobian matrix, we can synthesize a compliant mechanism with the desired number of DOFs to achieve the desired functionality and optimize the direct kinematics for higher performance.

4.4.2 Differences Between the C-stiffness and the Stiffness

The objective function for CPM topology optimization used in Ref. [15] is developed on the basis of Chen and Wang's formulation [5], which uses the C-stiffness. Here, we modify the formulation using a new stiffness evaluation process that calculates the stiffness based on its definition.

The formulation proposed by Chen and Wang is used for compliant mechanisms with a single input and a single output. In such a compliant system, the relationship between the forces (f_{in} and f_{out}) and the displacements (u_{in} and u_{out}) at input and output ports is described by the mechanism stiffness matrix $\mathbf{K}_m$ [37] shown in Eq. (4.32),

$$\begin{bmatrix} f_{in} \\ f_{out} \end{bmatrix} = \underbrace{\begin{bmatrix} k_{11} & k_{12} \\ k_{21} & k_{22} \end{bmatrix}}_{\mathbf{K}_m} \begin{bmatrix} u_{in} \\ u_{out} \end{bmatrix}. \tag{4.32}$$

The diagonal elements of the mechanism stiffness matrix $\mathbf{K}_m$ [37] comprise the C-stiffness [5]. As shown in Eq. (4.33), they incorporate the C-stiffness at the input and output ports (k_{11} and k_{22}) into the formulation to optimize the topology of a hinge-free compliant mechanism;

$$\min \; -\underbrace{e^{-(GA-GA^*)^2}}_{f} \underbrace{k_{11}k_{22}}_{S}, \tag{4.33}$$

where GA is the geometric advantage of the mechanism, GA^* is the desired geometric advantage, and f and S represent the kinematic and stiffness constraints, respectively.

According to Eq. (4.32), the physical meanings of k_{11} and k_{21} are the forces that should be applied at the input and output ports if $u_{in} = 1$ and $u_{out} = 0$ are expected. The input C-stiffness k_{11} describes part of the stiffness relationship between f_{in} and u_{in}, as does the output C-stiffness k_{22}.

In contrast, the stiffness based on its definition fully describes the force and displacement relationship for one specific DOF. To obtain the input and output stiffnesses of this simple compliant system, the input and output compliances are calculated using the physical meaning of compliance. Let $f_{in} = 1$ and $f_{out} = 0$ and solve Eq. (4.32). The resulting input displacement u_{in} is the input compliance

$$u_{in} = \frac{k_{22}}{k_{11}k_{22} - k_{12}k_{21}}. \tag{4.34}$$

Let $f_{in} = 0$ and $f_{out} = 1$, and solve Eq. (4.32). The resulting output displacement u_{out} is the output compliance

$$u_{out} = \frac{k_{11}}{k_{11}k_{22} - k_{12}k_{21}}. \tag{4.35}$$

The inverses of the input and output compliances are the input and output stiffnesses, respectively;

$$\begin{aligned} k_{in} &= u_{in}^{-1} = k_{11} - k_{12}k_{21}/k_{22} \\ k_{out} &= u_{out}^{-1} = k_{22} - k_{12}k_{21}/k_{11}. \end{aligned} \tag{4.36}$$

Comparing k_{11} and k_{22} with k_{in} and k_{out}, respectively, shows that the C-stiffness is part of the stiffness. Although the C-stiffness has been used to successfully evaluate the mechanism's stiffness in many design problems, some problems show that the stiffness is more reliable than the C-stiffness. Therefore, this section modifies the formulation of Jacobian-based topology optimization by replacing the C-stiffness with the stiffness.

4.4.3 Formulations Using the New Stiffness Evaluation Process

As shown in Eq. (4.33), the kinematic constraint f and the stiffness constraint S are two conflicting sub-objectives. On one hand, the mechanism should be soft enough to deform and deliver motion. On the other hand, it should be stiff enough to transmit forces to the mobile platform and bear the external force. A general formulation for the problem of optimizing a CPM's topology is as follows:

$$\begin{aligned} \min \quad & \zeta(\mathbf{X}) = -f^{\omega} S^{(1-\omega)} \\ \text{s.t.} \quad & V(\mathbf{X}) \le V_o \end{aligned}, \tag{4.37}$$

where ω $(0 < \omega < 1)$ is a weight indicating the relative significance of the kinematic constraint f, $V(\mathbf{X})$ is the volume fraction of the candidate topology, and V_o is the allowed volume fraction.

The kinematic constraint f has two different forms; one is selected based on the design problem. The first form of the kinematic constraint, f_1, is suitable for CPMs with kinematics simple enough to be predefined by designers. This form forces the Jacobian matrix of the CPM $\mathbf{J}$ to be close to a desired Jacobian matrix $\mathbf{J}^*$ by minimizing the differences between J_{ji} and J^*_{ji}, which are the elements of $\mathbf{J}$ and $\mathbf{J}^*$, respectively. As a result, the desired DOFs and kinematic properties of the CPM can be expressed by the desired Jacobian matrix $\mathbf{J}^*$,

$$\begin{aligned} & \max \ f_1 = e^{-\sum (J_{ji} - J^*_{ji})^2} \\ & j = 1, 2, \ldots, 3 \,\&\, i = 1, 2, \ldots, n. \end{aligned} \tag{4.38}$$

The second form of the kinematic constraint, f_2, is suitable for CPMs with complex kinematics. This form maximizes the motion in the desired DOFs J^d_j and suppresses the others (J^c_j);

$$\max \ f_2 = e^{-\sum J^c_j} \prod J^d_j, \tag{4.39}$$

where J_j evaluates the workspace of the jth DOF, which is the sum of the squares of the elements in the corresponding row vector, as shown in Eq. (4.40). J^d_j is the workspace of a desired DOF, and J^c_j is the workspace of a constrained DOF. The natural exponential function forces each J^c_j to be close to zero;

$$J_j = \sum_{i=1}^{n} J^2_{ji}. \tag{4.40}$$

In our previous study, [15], the stiffness constraint S is satisfied by maximizing the input and output C-stiffness of the CPM. This chapter uses the stiffness calculated based on the definition instead of the C-stiffness for the evaluation. Mathematically, the stiffness constraint is the product of the input stiffness k^i_{in} and the output stiffness k^i_{out},

$$\max \ S = \prod_{i=1}^{n} k^i_{in} k^i_{out}. \tag{4.41}$$

The calculation of the Jacobian matrix $\mathbf{J}$ and the input and output stiffnesses is addressed in the following sections.

4.4.3.1 Unification of the Units of the Rotational and Translational Variables

Since the units of the rotational and translational variables are different, it is unfair to compare the rotational and translational parameters directly during the optimization process. Therefore, the characteristic length [19, 45] l_c is introduced. Based on the characteristic length l_c, we can define the equivalent moment $\widetilde{M}$ and the equivalent rotational displacement $\widetilde{\theta}_z$ as follows:

$$\begin{aligned} \widetilde{M} &= M/l_c \\ \widetilde{\theta}_z &= \theta_z l_c. \end{aligned} \tag{4.42}$$

Each element J_{ji} of the Jacobian matrix is the ratio of the jth DOF of $\mathbf{U}_o$ and the ith input. Since all the inputs are translational, only the rotational displacement θ_z in $\mathbf{U}_o$ is transformed into an equivalent rotational displacement $\widetilde{\theta}_z$. Therefore, the elements in the third row of $\mathbf{J}$ are multiplied by l_c to obtain an equivalent Jacobian matrix $\widetilde{\mathbf{J}}$,

$$\widetilde{J}_{3i} = J_{3i} l_c, \quad i = 1, 2, \ldots, n. \tag{4.43}$$

For the input and output stiffnesses, the output stiffness related to rotation should be transformed into an equivalent rotational stiffness. The relationship among the applied moment M, the rotational stiffness k_θ, and the rotational displacement θ_z is as follows:

$$M = k_\theta \theta_z. \tag{4.44}$$

Solving for the moment M and rotational displacement θ_z in Eq. (4.42) and substitute them into Eq. (4.44) yields

$$\widetilde{M} l_c = k_\theta \widetilde{\theta}_z / l_c. \tag{4.45}$$

The equivalent rotational stiffness $\widetilde{k}_\theta$ is calculated as follows:

$$\widetilde{k}_\theta = \widetilde{M}/\widetilde{\theta}_z = k_\theta / l_c^2. \tag{4.46}$$

4.4.3.2 Input and Output Stiffnesses of CPMs

Using the mechanism stiffness matrix $\mathbf{K}_m$, we calculate the input and output stiffness of a CPM based on the physical meaning of compliance. As shown in Eq. (4.47), $n + 3$ unit load vectors ($\mathbf{F}^i_{in}$ and $\mathbf{F}^j_{out}$) are applied to the DOFs of $\mathbf{K}_m$ in sequence to obtain the corresponding displacement vectors $\mathbf{U}^i_{in}$ and $\mathbf{U}^j_{out}$;

$$\begin{aligned} \mathbf{F}^i_{in} &= \mathbf{K}_m \mathbf{U}^i_{in}, \quad i = 1, 2, \ldots, n \\ \mathbf{F}^j_{out} &= \mathbf{K}_m \mathbf{U}^j_{out}, \quad j = 1, 2, 3. \end{aligned} \tag{4.47}$$

The displacement where the unit load is applied is the compliance in this direction, which can be extracted from the corresponding displacement vector using the relevant unit load vector. The inverses of the compliances are the input and output stiffnesses of this CPM.

$$\begin{aligned} k_{in}^{i} &= \left((\mathbf{F}_{in}^{i})^{\mathrm{T}}\mathbf{U}_{in}^{i}\right)^{-1}, \quad i = 1, 2, \ldots, n \\ k_{out}^{j} &= \left((\mathbf{F}_{out}^{j})^{\mathrm{T}}\mathbf{U}_{out}^{j}\right)^{-1}, \quad j = 1, 2, 3. \end{aligned} \tag{4.48}$$

4.4.4 Sensitivity Analysis

The sensitivity of the objective function is determined by the sensitivities of J_{ji}, k_{in}^{i}, and k_{out}^{j}. The sensitivity of J_{ji} is determined by the sensitivity of $\mathbf{J}$, which is calculated as follows:

$$\frac{\partial \mathbf{J}}{\partial x^{e}} = \mathbf{K}_{22}^{-1}\frac{\partial \mathbf{K}_{22}}{\partial x^{e}}\mathbf{K}_{22}^{-1}\mathbf{K}_{21} - \mathbf{K}_{22}^{-1}\frac{\partial \mathbf{K}_{21}}{\partial x^{e}}. \tag{4.49}$$

Since $\mathbf{K}_{22}$ and $\mathbf{K}_{21}$ are parts of $\mathbf{K}_m$, the sensitivity of $\mathbf{J}$ is determined by the sensitivity of $\mathbf{K}_m$. According to Eq. (4.48), the sensitivity of k_{in}^{i} is calculated using Eq. (4.50), and the sensitivity of k_{out}^{j} obtained similarly;

$$\frac{\partial k_{in}^{i}}{\partial x^{e}} = -(\mathbf{F}_{in}^{i})^{\mathrm{T}}\frac{\partial \mathbf{U}_{in}^{i}}{\partial x^{e}}(k_{in}^{i})^{2}, \tag{4.50}$$

where the sensitivity of $\mathbf{U}_{in}^{i}$ is as follows:

$$\frac{\partial \mathbf{U}_{in}^{i}}{\partial x^{e}} = -\left(\mathbf{K}_{m}^{-1}\frac{\partial \mathbf{K}_{m}}{\partial x^{e}}\mathbf{K}_{m}^{-1}\right)\mathbf{F}_{in}^{i}. \tag{4.51}$$

By substituting Eq. (4.51) into Eq. (4.50), the sensitivity of k_{in}^{i} is rewritten as

$$\frac{\partial k_{in}^{i}}{\partial x^{e}} = \left((\mathbf{U}_{in}^{i})^{\mathrm{T}}\frac{\partial \mathbf{K}_{m}}{\partial x^{e}}\mathbf{U}_{in}^{i}\right)(k_{in}^{i})^{2}. \tag{4.52}$$

According to Eqs. (4.13) and (4.14), the sensitivity of $\mathbf{K}_m$ is determined by the sensitivity of $\mathbf{K}_{ao}^{i}$,

$$\frac{\partial \mathbf{K}_{ao}^{i}}{\partial x^{e}} = -\mathbf{K}_{ao}^{i}\left(\begin{bmatrix} 1 & \mathbf{0} \\ \mathbf{0} & \mathbf{T}_{i}^{o} \end{bmatrix} \otimes \frac{\partial \mathbf{C}_{ao}^{i}}{\partial x^{e}}\right)\mathbf{K}_{ao}^{i}, \tag{4.53}$$

where the sensitivity of an element of $\mathbf{C}_{ao}^{i}$ is calculated as follows:

$$
\begin{aligned}
\frac{\partial \mathbf{C}^i_{ao}(j,k)}{\partial x^e} &= \mathbf{F}_j^{\mathrm{T}} \frac{\partial \mathbf{U}_k}{\partial x^e} = \mathbf{F}_j^{\mathrm{T}} \frac{\partial}{\partial x^e}(\mathbf{K}_i^{-1}\mathbf{F}_k) \\
&= -\mathbf{F}_j^{\mathrm{T}} \mathbf{K}_i^{-1} \frac{\partial \mathbf{K}_i}{\partial x^e} \mathbf{K}_i^{-1} \mathbf{F}_k \\
&= -\mathbf{U}_j^{\mathrm{T}} \frac{\partial \mathbf{K}_i}{\partial x^e} \mathbf{U}_k \\
&= -\rho (x^e)^{\rho-1} \mathbf{U}_j^{\mathrm{T}} \mathbf{K}_e \mathbf{U}_k.
\end{aligned}
\tag{4.54}
$$

In the following sections, we compare several topology optimization results obtained using the C-stiffness and stiffness formulations. The artificial material properties for these examples are as follows: the Young's modulus is $E = 1\,GPa$ and the Poisson's ratio is $\upsilon = 0.3$. The characteristic length l_c is set to 10 mm in this study. All the numerical calculations are performed by a computer with an Intel Core $i7 - 6700$ (3.40 GHz) CPU, 8.00 GB of RAM and MATLAB $R2009a$. The filter length scale $d_{min} = 2.4$ is represented by a red bar in each figure showing the final topology.

For each numerical example in the next sections, the initial design $\mathbf{X}$ is defined by setting the material density of each element to the allowed volume fraction, i.e., $x^e = V_o$. The convergence criterion is based on the change in the design variables, which is set to 0.005. Note that the volume constraint is active throughout the optimization process.

4.4.5 *Design of a 2-DOF CPM*

In this section, a 2-DOF CPM is synthesized using two asymmetrical compliant limbs in the design domain shown in Fig. 4.19. As can be seen, the two compliant limbs have the same size and boundary conditions. Each compliant limb is discretized into 100×100 finite elements for elastic analysis. The amount of material allowed is 20%.

4.4.5.1 Solution Using f_1

Since the kinematics of a 2-DOF CPM is simple enough that predefine the desired Jacobian matrix $\mathbf{J}^*$ can be predefined, this design problem can be solved using the first form of the objective function in Eq. (4.38). The $\mathbf{J}^*$ for this example is given in Eq. (4.55), where a_{1x} and a_{2y} are the input DOFs. The zero vector in the third row of $\mathbf{J}^*$ means that θ_z is suppressed, i.e., the CPM is expected to have only two translational DOFs (u_x and u_y). Moreover, the two translational DOFs are expected to be decoupled. For example, the input a_{1x} only induces a translational displacement u_x with desired geometric advantage GA^* and has no impact on the other two DOFs, i.e.,

Fig. 4.19 Design domain for a 2-DOF CPM

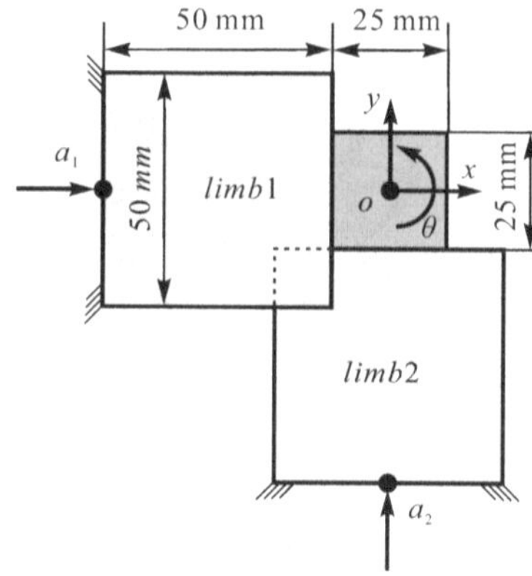

$$\begin{bmatrix} u_x \\ u_y \\ \theta_z \end{bmatrix} = \underbrace{\begin{bmatrix} GA^* & 0 \\ 0 & GA^* \\ 0 & 0 \end{bmatrix}}_{\mathbf{J}^*} \begin{bmatrix} a_{1x} \\ a_{2y} \end{bmatrix}. \tag{4.55}$$

First, GA^* is set to -3 and ω is set to 0.5. The corresponding topology optimization problem is solved using the C-stiffness and stiffness formulations. The optimization process is run for 200 iterations. The two resulting topologies are shown in Fig. 4.20. The C-stiffness and stiffness formulations both result in valid topologies in this case. Figures 4.21 and 4.22 show the iteration history of the objective function, the kinematic constraint f and the stiffness constraint S throughout the optimization process. The iteration curves oscillate. These oscillations may be caused by the distribution of material at specific elements. Fortunately, the topology becomes stable later in the iteration process; this configuration can be regarded as the optimal topology. The $\mathbf{J}$ corresponding to each of the two final topologies is listed in the first two rows of Table 4.2. Note that only the elements of the first column of $\mathbf{J}$ are displayed since $J_{11} \simeq J_{22}$, $J_{21} \simeq J_{12}$, and $J_{31} \simeq J_{32}$. One can see that the C-stiffness and stiffness formulations both force the $\mathbf{J}$ of the CPM to be close to $\mathbf{J}^*$; i.e., the kinematic constraint f is satisfied. In addition, the C-stiffness (Ck_{a1} and Ck_{ox}) and the stiffness (k_{a1} and k_{ox}) of the input and output ports on the x-axes of the two final topologies are given in the first two rows of Table 4.3. The results show that the C-stiffness and stiffness formulations both satisfy the stiffness constraint S effectively. The resulting output stiffness is smaller than the input stiffness required to produce the GA, whereas the C-stiffness is larger than the stiffness for the same mechanism.

However, when GA^* is positive, e.g., $GA^* = 2.5$, the C-stiffness formulation fails to satisfy the stiffness constraint S and results in an invalid topology. Figure 4.23a shows the final topology obtained using the C-stiffness with $GA^* = 2.5$. Although the input and output ports of the CPM are successfully connected by solid material, there is no material connection between the compliant limbs and the fixed ports. Consequently, the displacements of the input and output ports in one direction are equal,

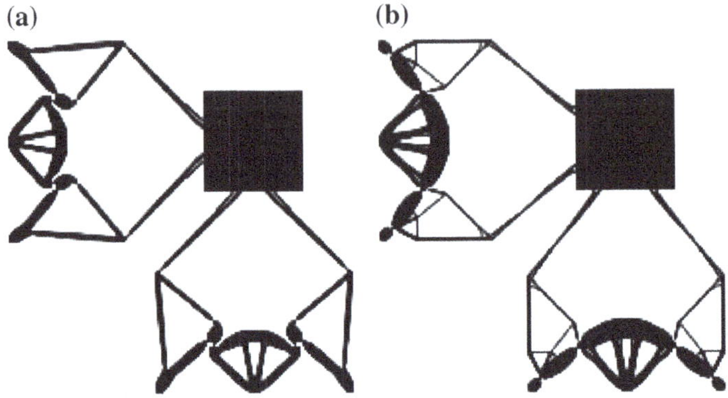

Fig. 4.20 Final topologies of the 2-DOF CPM for $GA^* = -3$ solved using **a** the C-stiffness formulation ($\omega = 0.5$) and **b** the stiffness formulation ($\omega = 0.5$)

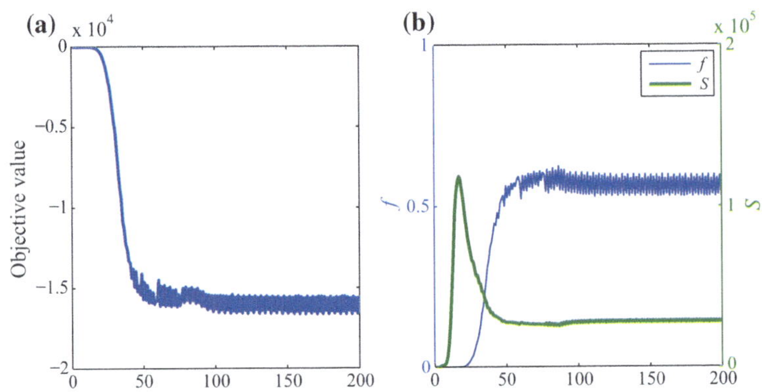

Fig. 4.21 Iteration history of the topology in Fig. 4.20a: **a** the objective function and **b** the kinematic constraint f and the stiffness constraint S

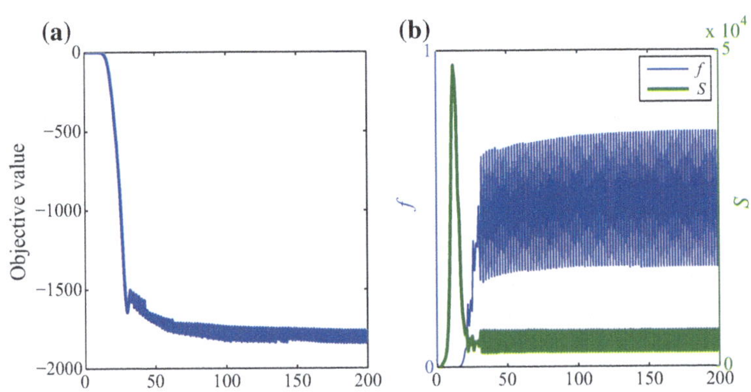

Fig. 4.22 Iteration history of the topology in Fig. 4.20b: **a** the objective function and **b** the kinematic constraint f and the stiffness constraint S

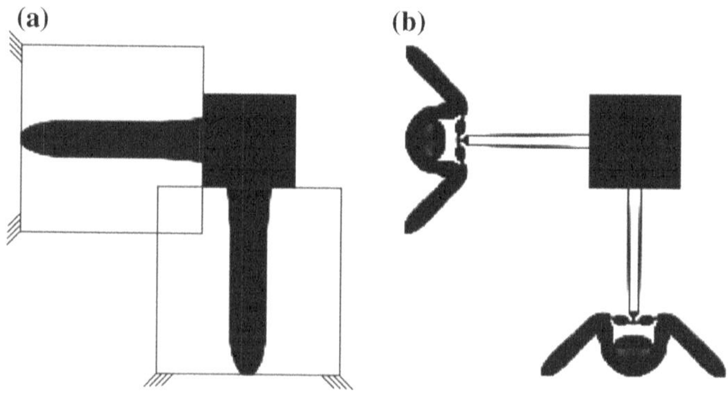

Fig. 4.23 Final topologies of a 2-DOF CPM with $GA^* = 2.5$ using **a** the C-stiffness formulation and **b** the stiffness formulation ($\omega = 0.7$)

Table 4.2 Jacobian matrices of the final 2-DOF CPMs

Formulation	GA^*	J_{11}	J_{21}	J_{31}
C-stiffness	−3	−2.52	8×10^{-4}	0.02
Stiffness	−3	−2.26	1.1×10^{-3}	0.02
C-stiffness	2.5	1	0	-3.5×10^{-3}
Stiffness	2.5	2.27	-4×10^{-4}	−0.02

Table 4.3 Input and output stiffnesses of the final 2-DOF CPMs

Formulation	GA^*	Ck_{a1}	Ck_{ox}	k_{a1}	k_{ox}
C-stiffness	−3	38.7	4.3	11.3	1.3
Stiffness	−3	51.4	3.1	35.8	2.1
C-stiffness	2.5	101.4	101.4	2×10^{-6}	2×10^{-6}
Stiffness	2.5	77.5	6.9	42.1	3.7

i.e., $J_{11} = 1$ (in the third row of Table 4.2). As shown in the third row of Table 4.3, the C-stiffnesses of the input and output ports on the x-axis are all 101.4 N/mm, whereas the corresponding stiffnesses k_{a1} and k_{ox} of this invalid topology are approximately zero. Obviously, the C-stiffness fails to evaluate the stiffness of the mechanism in this case.

This problem is resolved using the stiffness formulation. The final topology is shown in Fig. 4.23b, where ω is set to 0.7. One can see that there are valid material connections between the input, output and fixed ports of the CPM. As shown in the last rows of Tables 4.2 and 4.3, the resulting $\mathbf{J}$ is close to $\mathbf{J}^*$, e.g., $J_{11} = 2.27$. The values of its C-stiffness and stiffness are reasonable. Therefore, the problem with the C-stiffness is avoided and the validity of the final topology can be ensured using the stiffness formulation.

4.4.5.2 Solution Using f_2

To compare the two forms of the objective function, the 2-DOF CPM design problem is also solved using f_2, which is shown in Eq. (4.39). In this design problem, the kinematic constraint f_2 maximizes the workspace of the two translational DOFs and minimizes the workspace of the rotational DOF, which is formulated as follows:

$$\max\ f_2 = e^{-J_3^c} \prod_{j=1}^{2} J_j^d. \tag{4.56}$$

When f_2 is combined with the stiffness constraint S based on the C-stiffness, the resulting final topology is similar to the topology shown in Fig. 4.23a. The final topology obtained using f_2 and the stiffness-based version of S is shown in Fig. 4.24. ω is set to 0.7. The corresponding Jacobian matrix is described by Eq. (4.57). One can see that the workspaces of the two translational DOFs (u_x and u_y) are much larger than that of the rotational DOF θ_z. The resulting values of J_{11} and J_{22} are positive, i.e., f_2 does not control the sign of J_{ji} as f_1 does.

$$\mathbf{J} = \begin{bmatrix} 4.19 & 0.01 \\ -0.01 & 4.20 \\ -0.01 & 0.05 \end{bmatrix} \tag{4.57}$$

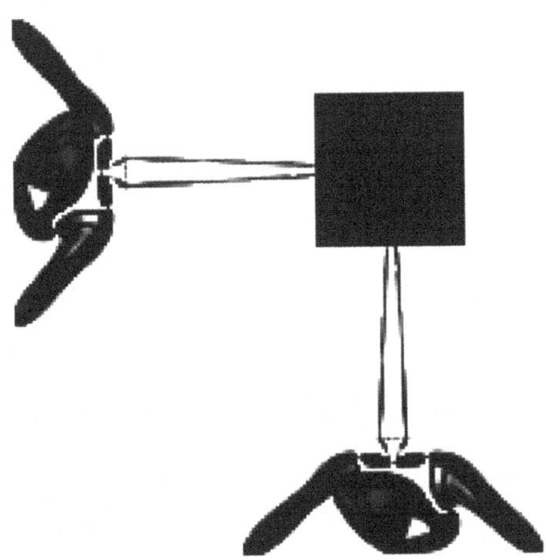

Fig. 4.24 Final topology of the 2-DOF CPM obtained using f_2 and the stiffness formulation

4.4.6 Design of a 3-DOF CPM

The second CPM design problem is to synthesize a 3-DOF CPM with three symmetrically arranged compliant limbs. The position of the CPM's input, output and fixed points are shown in Fig. 4.25. Each design domain Ω_i is discretized into 50×80 finite elements for elastic analysis with the same boundary conditions. The amount of material allowed is 20%. Since a 3-DOF CPM has complex kinematics, it is difficult to predefine a desired Jacobian matrix for the optimization process. The second form of the kinematic constraint, f_2, shown in Eq. (4.39), is used with the objective function. In a planar 3-DOF CPM, no degree of freedom is suppressed, i.e., the design objective is to maximize the workspaces of the three DOFs.

The problem is solved using the C-stiffness and stiffness formulations. The optimization process is run for 200 iterations. Figure 4.26 shows the final topology of the 3-DOF CPM obtained using the C-stiffness formulation with $\omega = 0.7$. Its corresponding Jacobian matrix **J** is shown in Eq. (4.58). The final topology obtained using the stiffness formulation ($\omega = 0.9$) and its resulting Jacobian matrix **J** are shown in Fig. 4.27 and Eq. (4.59), respectively. One can see that the C-stiffness and stiffness formulations both result in valid topologies in this example, but the topology obtained using the stiffness formulation is stiffer.

$$\mathbf{J} = \begin{bmatrix} 0.22 & -1.79 & 1.57 \\ 1.94 & -0.78 & -1.16 \\ 0.27 & 0.27 & 0.27 \end{bmatrix} \tag{4.58}$$

$$\mathbf{J} = \begin{bmatrix} 0.93 & -2.18 & 1.24 \\ 1.97 & -0.18 & -1.79 \\ 0.12 & 0.12 & 0.12 \end{bmatrix} \tag{4.59}$$

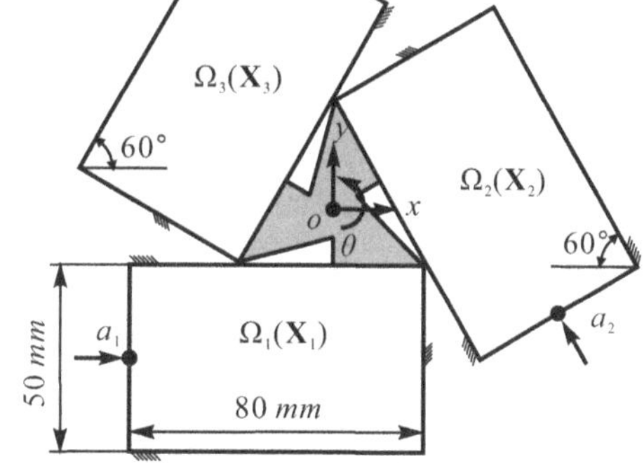

Fig. 4.25 Design domain for a 3-DOF CPM

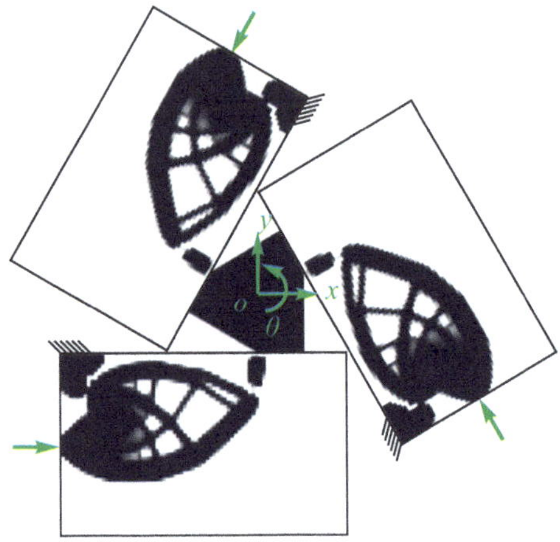

Fig. 4.26 Final topology of the 3-DOF CPM using the C-stiffness formulation

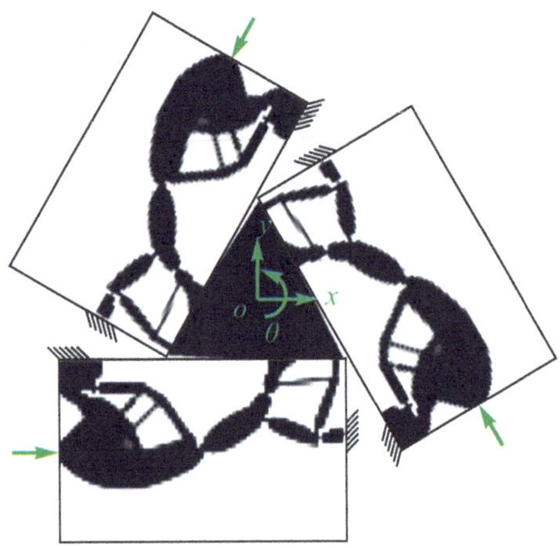

Fig. 4.27 Final topology of the 3-DOF CPM using the stiffness formulation

4.4.7 Solving Benchmark Problems

Although the Jacobian-based topology optimization method was developed for CPMs, this method is applicable to typical compliant mechanisms designed using current topology optimization methods. Here, we take consider a displacement inverter as an example to illustrate this statement.

4.4.7.1 Displacement Inverter and Amplifier

The problem of designing a 1-DOF compliant mechanism using a single design domain is shown in Fig. 4.28. The top left corner and the bottom left corner of the design domain are fixed. The input point a and the output point o are centered on the

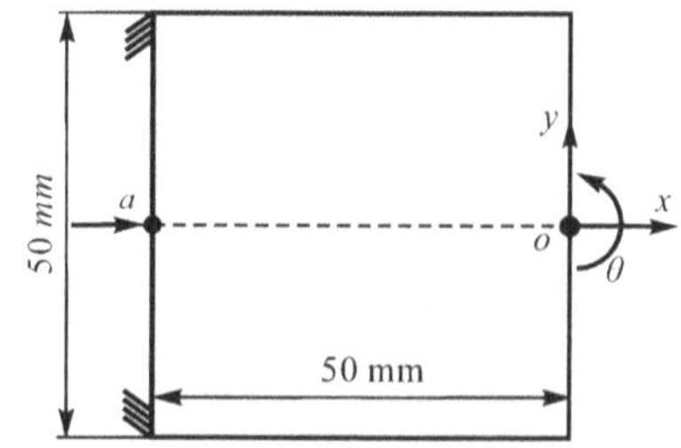

Fig. 4.28 Design domain for a displacement inverter and amplifier

left and right sides, respectively. The design domain is discretized into 100 × 100 finite elements for elastic analysis. The material usage is restricted to 20%. The Jacobian matrix of the 1-DOF compliant mechanism is a 3 × 1 vector whose desired form is shown in Eq. (4.60). One can see that the desired output motion of this mechanism is along the x-axis, and the DOFs u_y and θ_z should be suppressed. Obviously, when only element J_{11} of the Jacobian matrix is considered, our objective function is equal to one proposed by Chen and Wang Eq. (4.33),

$$[u_x, u_y, \theta_z]^{\mathrm{T}} = [GA^*, 0, 0]^{\mathrm{T}} a_x. \tag{4.60}$$

When GA^* is negative, e.g., $GA^* = -3$, the design problem is to synthesize a displacement inverter. The corresponding topology optimization problem is solved using the C-stiffness and stiffness formulations. Figure 4.29a shows the topology obtained by using the C-stiffness formulation and with $\omega = 0.5$. Its corresponding Jacobian matrix is $\mathbf{J} = [-2.7, 0, 0]^{\mathrm{T}}$, i.e., the optimized GA of the displacement inverter is -2.7. The C-stiffnesses at the input and output ports are 40.9 and 4.7 N/mm, respectively, while the stiffnesses at the input and output ports are 6.4 and 0.7 N/mm, respectively. Figure 4.29b shows the topology obtained using the stiffness formulation and with ω set to 0.5. The corresponding Jacobian matrix is $\mathbf{J} = [-2.8, 0, 0]^{\mathrm{T}}$. The C-stiffnesses at the input and output ports are 57.2 and 2.7 N/mm, respectively, while the stiffnesses at input and output ports are 41.3 and 2.0 N/mm, respectively. One can see that the C-stiffness and stiffness formulations both produce valid topologies in this case.

When GA^* is positive, e.g., $GA^* = 3$, the design problem is to synthesize a displacement amplifier. The C-stiffness formulation results in the invalid topology shown in Fig. 4.30a. Without material connected to the fixed ports, the Jacobian matrix of the resulting displacement amplifier is $\mathbf{J} = [1, 0, 0]^{\mathrm{T}}$. The input and output C-stiffnesses are both 116.1 N/mm, and the corresponding input and output stiffnesses are both 4.3×10^{-7} N/mm. Figure 4.30b shows the topology obtained using the stiffness formulation and with ω set to 0.6. The corresponding Jacobian matrix is $\mathbf{J} = [2.3, 0, 0]^{\mathrm{T}}$. The C-stiffnesses at the input and output ports are 79.5 and 6.2 N/mm, respectively, while the stiffnesses at the input and output ports are 47.8 and 3.8 N/mm, respectively. This case also shows that the C-stiffness formulation fails to satisfy the stiffness constraint.

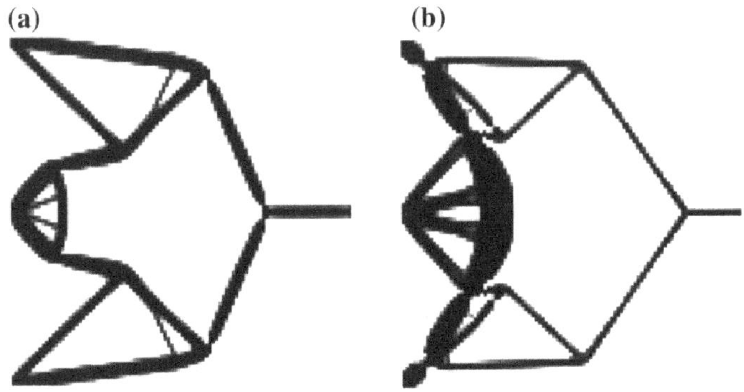

Fig. 4.29 Final topologies of a displacement inverter for $GA^* = -3$ solved using **a** the C-stiffness formulation ($\omega = 0.5$) and **b** the stiffness formulation ($\omega = 0.5$)

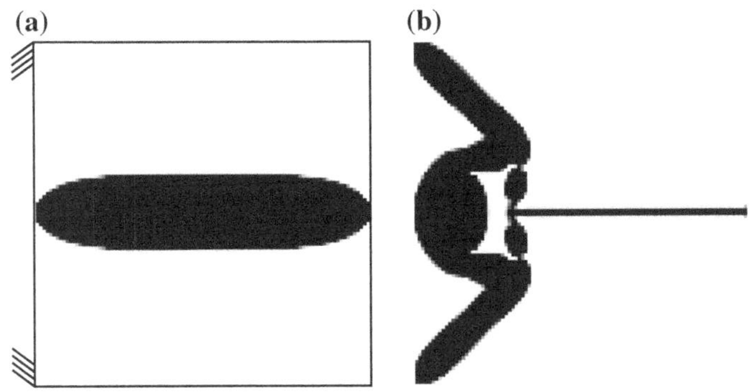

Fig. 4.30 Final topologies of a displacement amplifier for $GA^* = 3$ solved using **a** the C-stiffness formulation and **b** the stiffness formulation ($\omega = 0.6$)

4.4.7.2 Solution Using the Artificial I/O Spring and MSE/SE Formulations

Several popular formulations of the topology optimization of compliant mechanisms have been developed. Deepak et al. [6] have compared these formulations. The popular artificial I/O spring and MSE/SE (mutual strain energy/strain energy) formulations are adopted here to solve the problem of a displacement amplifier. In the artificial I/O spring formulation, artificial springs are added to the input and output ports. The problem is formulated as follows:

$$\begin{aligned} \max \quad & u_{out} \\ \text{s.t.} \quad & V(\mathbf{X}) \leq V_o. \end{aligned} \tag{4.61}$$

For the problem of a displacement amplifier discussed above, the artificial I/O spring formulation results in the invalid topology shown in Fig. 4.30a, but the

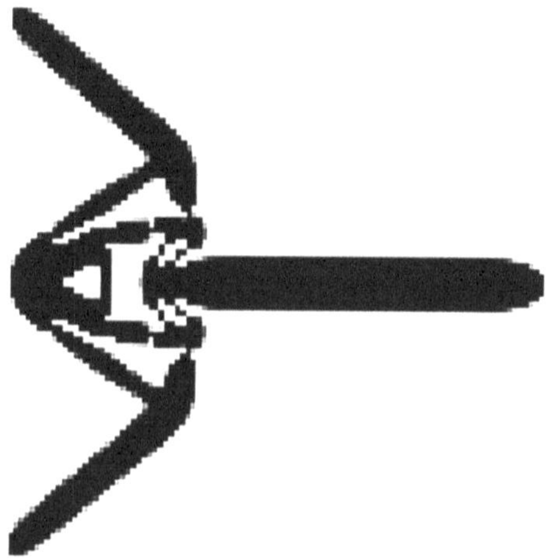

Fig. 4.31 Final topology of the displacement amplifier designed using the MSE/SE formulation

stiffnesses of the artificial springs vary from 10^{-10} to 10^{10} N/mm. In contrast, the MSE/SE formulation Eq. (4.62) produces the valid topology shown in Fig. 4.31.

$$\begin{aligned} \min \quad & -MSE + SE \\ \text{s.t.} \quad & V(\mathbf{X}) \le V_o \end{aligned} \tag{4.62}$$

The computational expense of the proposed method is compared with that of the artificial I/O spring formulation by designing a displacement inverter in Sect. 5.3.1. It takes the artificial I/O spring formulation 47.59 s and 220 iterations to reach the solution, i.e., 216 ms per iteration. The C-stiffness formulation requires 181.5 s and 300 iterations to obtain the topology in Fig. 4.29a, i.e., 605 ms per iteration. The stiffness formulation requires 209.37 s and 300 iterations to obtain the topology in Fig. 4.29b, i.e., 698 ms per iteration. Obviously, the proposed method is more expensive than the artificial I/O spring formulation in computation. One reason for this is that using finite elements with 12 nodal DOFs increases the computational cost of FEA. The other reason is that all the DOFs of the output point o are considered in the proposed method, whereas the spring formulation is only concerned with u_x.

4.5 Conclusions

In this chapter, the status of research into CPM synthesis is introduced; there is currently no method for synthesizing CPMs that considers topology performance. Traditional topology optimization cannot be applied to CPMs directly because the method of predefining multiple specific ports is too complicated for multi-DOF CPMs. Jacobian-based topology optimization adopts multiple design domains to synthesize a CPM's compliant limbs based on the performance of the complete

CPM. FEA and the matrix method are combined to build a new analysis frame for multi-DOF compliant mechanisms. On this basis, a different objective formulation is proposed. The design examples show that the CPM's function and kinematic performance can be optimized simultaneously by optimizing its Jacobian matrix. Maximizing the characteristic input and output stiffnesses guarantees a valid material connection during topology optimization. Additionally, the applicability of the Jacobian-based topology optimization method is verified by solving benchmark problems.

References

1. Ansola, R., Veguería, E., Maturana, A., Canales, J.: 3D compliant mechanisms synthesis by a finite element addition procedure. Finite Elem. Anal. Des. **46**(9), 760–769 (2010)
2. Bendsøe, M.P., Sigmund, O.: Topology Optimization: Theory, Methods and Applications. Springer, Heidelberg (2004)
3. Bendsoe, M.P., Sigmund, O.: Topology Optimization: Theory, Methods, and Applications. Springer Science & Business Media (2013)
4. Kim, C.J., S.K., Moon, Y.M.: An instant center approach toward the conceptual design of compliant mechanisms. ASME J. Mech. Des. **128**(3), 542–550 (2006)
5. Chen, S., Wang, M.Y.: Designing distributed compliant mechanisms with characteristic stiffness. In: ASME 2007 IDETC/CIE, pp. DETC2007–34,437. ASME, Las Vegas, NV, USA (2007)
6. Deepak, S.R., Dinesh, M., Sahu, D.K., Ananthasuresh, G.: A comparative study of the formulations and benchmark problems for the topology optimization of compliant. J. Mech. Robot. **1**(1), 1–8 (2009)
7. Frecker, M., Kikuchi, N., Kota, S.: Topology optimization of compliant mechanisms with multiple outputs. Struct. Multidiscip. Optim. **17**(4), 269–278 (1999)
8. Frecker, M.I., Ananthasuresh, G.K., Nishiwaki, S., Kikuchi, N., Kota, S.: Topological synthesis of compliant mechanisms using multi-criteria optimization. ASME J. Mech. Des. **119**(2), 238–245 (1997)
9. Herve, J.M.: Analyse structurelle des mécanismes par groupe des déplacements. Mech. Mach. Theory **13**(4), 437–450 (1978)
10. Hopkins, J.B., Culpepper, M.L.: Synthesis of multi-degree of freedom, parallel flexure system concepts via freedom and constraint topology (fact)-part I: principles. Precis. Eng. **34**(2), 259–270 (2010)
11. Howell, L.L.: Compliant Mechanisms. Wiley (2001)
12. Huang, Z., Li, Q.: General methodology for type synthesis of symmetrical lower-mobility parallel manipulators and several novel manipulators. Int. J. Robot. Res. **21**(2), 131–145 (2002)
13. Hudgens, J., Tesar, D.: A fully-parallel six degree-of-freedom micromanipulator: kinematic analysis and dynamic model. In: Proceedings of 20th Biennial ASME Mechanisms Conference, vol. 15-3, pp. 29–37. ASME, Kissimmee, FL, USA (1988)
14. Jang, G.W., Lee, S., Kim, Y.Y., Sheen, D.: Topology optimization using non-conforming finite elements: three-dimensional case. Int. J. Numer. Meth. Eng. **63**(6), 859–875 (2005)
15. Jin, M., Zhang, X.: A new topology optimization method for planar compliant parallel mechanisms. Mech. Mach. Theory **95**, 42–58 (2016)
16. Jin, M., Zhang, X., Zhu, B.: Design of compliant mechanisms using a pseudo-rigid-body model based topology optimization method. In: ASME 2014 IDETC/CIE, pp. DETC2014–34,325. ASME, Buffalo, USA (2014)

17. Jin, M., Zhang, X., Zhu, B.: A numerical method for static analysis of pseudo-rigid-body model of compliant mechanisms. Proc. Inst. Mech. Eng. Part C J. Mech. Eng. Sci. **228**(17), 3170–3177 (2014)
18. Jin, M., Zhang, X., Zhu, B., Wang, N.: Spring-joint method for topology optimization of planar passive compliant mechanisms. Chin. J. Mech. Eng. **26**(6), 1063–1072 (2013)
19. Kim, C.J., Moon, Y.M., Kota, S.: A building block approach to the conceptual synthesis of compliant mechanisms utilizing compliance and stiffness ellipsoids. J. Mech. Des. **130**(2), 284–284 (2008)
20. Kim, H., Kim, J., Ahn, D., Gweon, D.: Development of a nanoprecision 3-DOF vertical positioning system with a flexure hinge. IEEE Trans. Nanotechnol. **12**(2), 234–245 (2013)
21. Kong, X., Gosselin, C.M.: Type synthesis of 3-DOF translational parallel manipulators based on screw theory. ASME J. Mech. Des. **126**(1), 83–92 (2004)
22. Koseki, Y., Tanikawa, T., Koyachi, N., Arai, T.: Kinematic analysis of translational 3-DOF micro parallel mechanism using matrix method. In: Proceedings of the 2000 IEEE/RSJ International Conference on Intelligent Robots and Systems, vol. 1, pp. 786–792 (2000)
23. Li, Y., Xu, Q.: Design and analysis of a totally decoupled flexure-based xy parallel micromanipulator. IEEE Trans. Robot. **25**(3), 645–657 (2009)
24. Lu, K.J., Kota, S.: An effective method of synthesizing compliant adaptive structures using load path representation. J. Intell. Mater. Syst. Struct. **16**(4), 307–317 (2005)
25. Lu, K.J., Kota, S.: Topology and dimensional synthesis of compliant mechanisms using discrete optimization. J. Mech. Des. **128**(5), 1080–1091 (2006)
26. Lum, G.Z., Teo, T.J., Yang, G., Yeo, S.H., Sitti, M.: A hybrid topological and structural optimization method to design a 3-DOF planar motion compliant mechanism. In: 2013 IEEE/ASME International Conference on Advanced Intelligent Mechatronics (AIM), pp. 247–254. IEEE, Wollongong, Australia (2013)
27. Lum, G.Z., Teo, T.J., Yang, G., Yeo, S.H., Sitti, M.: Integrating mechanism synthesis and topological optimization technique for stiffness-oriented design of a three degrees-of-freedom flexure-based parallel mechanism. Precis. Eng. **39**, 125–133 (2015)
28. McInroy, J.E., Hamann, J.C.: Design and control of flexure jointed hexapods. IEEE Trans. Robot. Autom. **16**(4), 372–381 (2000)
29. Pucheta, M.A., Cardona, A.: Design of bistable compliant mechanisms using precision-position and rigid-body replacement methods. Mech. Mach. Theory **45**(2), 304–326 (2010)
30. Putra, A.S., Huang, S., Tan, K.K., Panda, S.K., Lee, T.H.: Design, modeling, and control of piezoelectric actuators for intracytoplasmic sperm injection. IEEE Trans. Control Syst. Technol. **15**(5), 879–890 (2007)
31. Saxena, A.: Topology design of large displacement compliant mechanisms with multiple materials and multiple output ports. Structural Multidisciplinary Optimization **30**(6), 477–490 (2005)
32. Saxena, A., Ananthasuresh, G.K.: A computational approach to the number of synthesis of linkages. ASME J. Mech. Des. **125**(1), 110–118 (2003)
33. Sedlaczek, K., Gaugele, T., Eberhard, P.: Topology optimized synthesis of planar kinematic rigid body mechanisms. Multibody Dynamics **137**(7), 251–260 (2006)
34. Sigmund, O.: Design of multiphysics actuators using topology optimization-part I: one-material structures. Comput. Methods Appl. Mech. Eng. **190**(49), 6577–6604 (2001)
35. Wadikhaye, S., Yong, Y., Moheimani, S.: Design of a compact serial-kinematic scanner for high-speed atomic force microscopy: an analytical approach. IET Micro Nano Lett. **7**(4), 309–313 (2012)
36. Wang, H., Zhang, X.: Input coupling analysis and optimal design of a 3-DOF compliant micro-positioning stage. Mech. Mach. Theory **43**(4), 400–410 (2008)
37. Wang, M.Y.: A kinetoelastic formulation of compliant mechanism optimization. J. Mech. Robot. **1**(2), 021,011 (2009)
38. Wang, R., Zhang, X.: Optimal design of a planar parallel 3-DOF nanopositioner with multi-objective. Mech. Mach. Theory **112**, 61–83 (2017)

39. Yamada, T., Izui, K., Nishiwaki, S., et al.: A topology optimization method based on the level set method incorporating a fictitious interface energy. Comput. Methods Appl. Mech. Eng. **199**(45), 2876–2891 (2010)
40. Yang, T.L., Liu, A.X., Yao, F.H., Jin, Q., Luo, Y.F.: Structure synthesis of 4-DOF (3-translation and 1-rotation) parallel robot mechanisms based on the units of single-opened-chain. In: Proceedings of the ASME Design Engineering Technical Conference, vol. 2, pp. 1295–1302. Pittsburgh (2001)
41. Yi, B.J., Chung, G.B., Na, H.Y., Kim, W.K., Suh, I.H.: Design and experiment of a 3-DOF parallel micromechanism utilizing flexure hinges. IEEE Trans. Robot. Autom. **19**(4), 604–612 (2003)
42. Yong, Y.K., Lu, T.F.: Kinetostatic modeling of 3-RRR compliant micro-motion stages with flexure hinges. Mech. Mach. Theory **44**(6), 1156–1175 (2009)
43. Yu, J., Li, S., Su, H.j., Culpepper, M.: Screw theory based methodology for the deterministic type synthesis of flexure mechanisms. J. Mech. Robot. **3**(3), 031,008 (2011)
44. Zhang, X., Ouyang, G., Wang, H.: Topology optimization of multiple inputs and multiple outputs compliant mechanisms. Chin. J. Mech. Eng. **20**(1), 82–85 (2007)
45. Zhang, Y., Su, H., Liao, Q.: Mobility criteria of compliant mechanisms based on decomposition of compliance matrices. Mech. Mach. Theory **79**, 80–93 (2014)
46. Zhu, B., Zhang, X., Wang, N., Fatikow, S.: Topology optimization of hinge-free compliant mechanisms using level set methods. Eng. Optimization **46**(5), 580–605 (2014)

Chapter 5
Extensions

Abstract In this chapter, we discuss some extensions to topology optimization methods for designing compliant mechanisms. The computational efficiency of level set-based topology optimization methods and geometric nonlinearity-, reliability- and multimaterial-based design problems, are discussed in detail. Some benchmark design problems are used to demonstrate the validity of the proposed methods.

5.1 Improving the Computational Efficiency of the Level Set Method

In standard level set-based topology optimization methods, the structural boundary is treated as a design parameter and is implicitly represented as the zero level set of a higher-dimensional scalar function. The optimum configuration is identified by solving the so-called Hamilton-Jacobi equation in an Eulerian coordinate system.

However, because of the CFL condition, the time step for updating the level set function is determined by the smallest grid, which is generally very small. Therefore, many iterations are often required to obtain an accurate optimal solution [6, 19]. Too many iterations make the optimization process computationally expensive and significantly decrease the numerical stability due to the accumulation of truncation errors [24].

To improve the computational efficiency, a velocity mapping algorithm was developed by Mei and Wang [14] to speed up the evolution of the level set equation. Wang et al. [20] proposed two numerical techniques, velocity mapping and variational regularization, for improving the computational speed. Luo et al. [11] developed a semi-implicit level set method for structural topology optimization that allows enhanced relaxation of the time step size, which is originally limited by the CFL condition of the explicit schemes. Thus, the computational efficiency is significantly improved. Recently, Zhou and Wang [24] proposed a semi-Lagrangian level set method for structural optimization. In their method, a line search algorithm and a sensitivity modulation scheme are introduced to the semi-Lagrangian method to improve the overall computational efficiency by requiring fewer design iterations. A level set method without re-initialization that was developed by Zhu and Zhang [25] uses

X. Zhang and B. Zhu, *Topology Optimization of Compliant Mechanisms*,
https://doi.org/10.1007/978-981-13-0432-3_5

an energy functional to ensure that the level set function remains a signed distance function (SDF) and avoid the need for re-initialization.

5.1.1 A Velocity Predictor-Corrector Construction Scheme

The optimization process is transferred into the evolution of the level set function ϕ at time t with a normal velocity V_n. During the optimization process, the gradients of the level set function at the design interface need to be controlled to decrease the numerical error and ensure stability. One method is to ensure that ϕ remains an SDF, i.e., $|\nabla\phi| = 1$, through re-initialization schemes. Therefore, the level set equation can be written as

$$\frac{\partial\phi}{\partial t} - V_n = 0 \tag{5.1}$$

and approximately rewritten as

$$\phi^{t+\Delta t} = \phi^t + V_n \Delta t, \tag{5.2}$$

where Δt denotes the time-step size and should satisfy

$$\Delta t \leq \frac{\min(\Delta x, \Delta y)}{\max(|V_n|)}, \tag{5.3}$$

where Δx and Δy are the grid spaces in the horizontal and vertical directions, respectively.

Suppose that in the kth iteration, the shape and topology of the structure are presented by ϕ_k. In the first two steps, the velocity fields V_{n1} and V_{n2} are obtained using the shape sensitivity analysis. Thus, the following equations are obtained:

$$\phi_2 = \phi_1 + V_{n1}\Delta t \tag{5.4}$$

$$\phi_3 = \phi_2 + V_{n2}\Delta t. \tag{5.5}$$

Suppose that $L(\phi_3)$, $L(\phi_2)$ and $L(\phi_1)$ are the Lagrangian values of the third, second and first iterations, respectively. Therefore, $L(\phi_3) < L(\phi_2) < L(\phi_1)$. V_{n1} and V_{n2}, respectively, can be expressed as

$$V_{n1} = -\left.\frac{\partial L(\phi)}{\partial\phi}\right|_{\phi=\phi_1} \tag{5.6}$$

$$V_{n2} = -\left.\frac{\partial L(\phi)}{\partial\phi}\right|_{\phi=\phi_2}. \tag{5.7}$$

In the third step, the velocity field, V_{n3}, is constructed using the following equation:

$$V_{n3} = \alpha V_{n2} + (1 - \alpha) V_{n1}, \tag{5.8}$$

where $0 < \alpha < 1$ is a constant and determined using

$$\alpha = \frac{V_{n2}^T V_{n1}}{1 + V_{n2}^T V_{n1}}. \tag{5.9}$$

Assume that V_{n1} and V_{n2} are obtained using the shape sensitivity analysis and that each guarantees a decrease in the Lagrangian L. The velocity field V_{n3} obtained using Eqs. (5.8) and (5.9) guarantees a decrease in the Lagrangian L. Even though $V_{n3} \neq -\left.\frac{\partial L(\phi)}{\partial \phi}\right|_{\phi=\phi_3}$, the choice of α according to Eq. (5.9) guarantees a decrease in the Lagrangian $L(\phi_5)$ compared to $L(\phi_4)$, where

$$\phi_5 = \phi_4 + V_{n4} \Delta t \tag{5.10}$$

$$V_{n4} = \alpha V_{n3} + (1 - \alpha) V_{n2} \tag{5.11}$$

$$\alpha = \frac{V_{n3}^T V_{n2}}{1 + V_{n3}^T V_{n2}}. \tag{5.12}$$

See Appendix A.1 for the proof of these statements.

5.1.2 *Optimization Algorithm*

A possible algorithm for improving the computational efficiency is shown in Algorithm 1.

Algorithm 1

1: Make an initial guess ϕ_1
2: Perform a shape sensitivity analysis to obtain V_{n1};
  $\phi_2 = \phi_1 + V_{n1} \Delta t$;
  Re-initialization
3: Perform a shape sensitivity analysis to obtain V_{n2};
  $\phi_3 = \phi_2 + V_{n2} \Delta t$;
  Re-initialization
4: **while** Not converged **do**
5:   $V_{n3} = \alpha V_{n2} + (1 - \alpha) V_{n1}$
6:   $\phi_4 = \phi_3 + V_{n3} \Delta t$
7:   Re-initialization
8:   $V_{n1} \Leftarrow V_{n2}$
9:   $V_{n2} \Leftarrow V_{n3}$
10: **end while**

Algorithm 2

1: Make an initial guess ϕ_1; select r_f
2: Perform a shape sensitivity analysis to obtain V_{n1};
   $\phi_2 = \phi_1 + V_{n1}\Delta t$;
   Re-initialization
3: Perform a shape sensitivity analysis to obtain V_{n2};
   $\phi_3 = \phi_2 + V_{n2}\Delta t$;
   Re-initialization
4: Convergence check
5: $k \Leftarrow 1$
6: **if** $k < r_f$ **then**
7: Calculate $V_{n2}^T V_{n1}$
8: **if** $V_{n2}^T V_{n1} \leq 0$ **then**
9: Break and go to step 2
10: **end if**
11: $\alpha = \frac{V_{n2}^T V_{n1}}{1+V_{n2}^T V_{n1}}$
12: $V_{n3} = \alpha V_{n2} + (1-\alpha) V_{n1}$
13: $\phi_4 = \phi_3 + V_{n3}\Delta t$
14: Re-initialization
15: $V_{n1} \Leftarrow V_{n2}$
16: $V_{n2} \Leftarrow V_{n3}$
17: $k++$
18: **else**
19: Go to step 3
20: **end if**

Nonetheless, due to the infinite-dimensional nature of the problem, the new velocity field generated using Eq. (5.8) may become less significant to the optimization process after a few iterations. If only the first two velocity fields are generated from the shape sensitivity and the FEA and all other velocity fields are based on the first two velocity fields, then, it is generally impossible to find the optimal solution because the first two velocity fields are closely related to the initial guess. Furthermore, after a few iterations, the two velocity fields may have $V_{n2}^T V_{n1} \leq 0$, which means that the new velocity field does not guarantee a decrease in the Lagrangian L.

It is advisable to renew the velocity field V_n based on the shape sensitivity analysis after a certain number of iterations to ensure that the computation remains meaningful and efficient; this is restarting. A flow chart demonstrating the proposed method is shown in Algorithm 2. Note that steps 6–20 are called the inner loop. The following two reset conditions are suggested:

$$V_{n2}^T V_{n1} \leq 0 \tag{5.13}$$

$$k = r_f, \tag{5.14}$$

where r_f determines how often the algorithm is restarted. If any one of the conditions occurs, then, the velocity is reset according to the shape sensitivity analysis.

The restart frequency r_f may depend on the objective function. However, any objective function calculation involves FEA and slows the optimization process. According to the analysis performed in the previous section, any $r_f > 0$ can be used to speed up the optimization process. However, when r_f is large, oscillations may occur during the optimization process. Based the authors' experience, r_f can be constrained using

$$0 < r_f \leq \min(nelx, nely), \tag{5.15}$$

where $nelx$ and $nely$ are the numbers of elements in the horizontal and vertical directions, respectively. Note that if $r_f = \min(nelx, nely)$ ensures the stability of the optimization process, any frequency less than $\min(nelx, nely)$ can be implemented. The reason for choosing $\min(nelx, nely)$ as the upper limit of r_f is as follows: To implement Eq. (5.8), suppose that all the obtained V_n are the same. Then, it takes at least $\min(nelx, nely)$ steps to move a boundary from one side of the design domain to the other side.

5.1.3 Examples

Two examples are presented to demonstrate the validity of the proposed method. The artificial material properties are as follows: the Young's modulus of the solid material is $E = 1$ and the Poisson's ratio is $\upsilon = 0.3$. The void area is assumed to have a Young's modulus $E = 0.001$ and the same Poisson's ratio $\upsilon = 0.3$. A computer with an Intel Core Due P8400 (2.26 GHz) CPU, 2.00 GB of RAM and MATLAB R2009b was used.

In accordance with the convergence criterion proposed in [3], convergence is achieved when the volume is within 0.005 of the required value Vol_{max} and the values of the objective function in the previous 5 steps are also within 1% of the values in current step.

5.1.3.1 Mean Compliance Minimization Problem

The design domain of the problem is shown in Fig. 5.1. The ratio of the length and height of design domain is 2 : 1. The bottom left and right corners are fixed. A single vertical load $F = 1$ is applied at the center of the bottom of the design domain. A maximum material usage constraint of 0.2 is considered. The design domain is discretized by 80×40 finite elements for the elastic analysis.

The computation time required to solve the bridge problem using four topology optimization approaches, namely the proposed method, the standard level set method [19], the discrete level set method [3] and the SIMP method [17], is compared. The initial configuration shown in Fig. 5.1 is used for both the standard and the proposed level set methods.

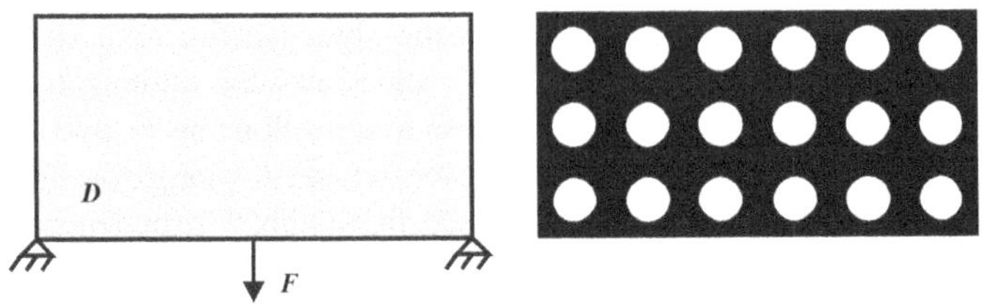

Fig. 5.1 The design domain for the bridge topology optimization problem (left) and the initial design used in both the proposed method and TOPLSM 199 (right)

In the implementation of the proposed method, $r_f = \min(nelx, nely)$ is used. In the implementation of the standard level set method, the MATLAB program TOPLSM 199 is used; it can be directly downloaded from http://www2.mae.cuhk.edu.hk/~cmdl/download.htm. The input command is TOPLSM(2, 1, 80, 40, 19.6, 0, 20, 5, 500). In the implementation of the SIMP method, the 99-line MATLAB program [17] is used, and the input command is Top(80, 40, 0.2, 3, 1.5). In the implementation of the discrete level set method, the code proposed in [3] is used, and the input command is top_levelset(80, 40, 0.2, 3, 2, 2).

The final designs obtained using those four methods are shown in Fig. 5.2. Nearly identical solutions are obtained, which confirms the ability of the proposed method to produce optimized solutions.

The computation time, final compliance and converged volume fraction are compared in Table 5.1. The convergence histories for the proposed and discrete level set methods are shown in Fig. 5.3. A total of 180 iterations is required for convergence when the standard level set method is used. With our method, fewer than 30 iterations are needed. The number of iterations is decreased by a factor of approximately 6.

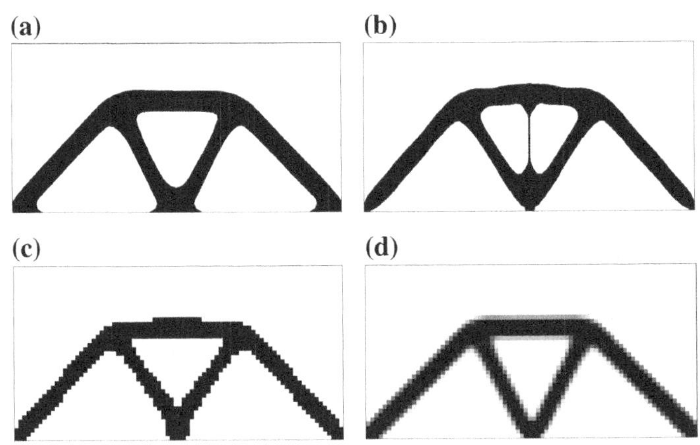

Fig. 5.2 The final solutions to the bridge problem obtained using **a** the proposed method, **b** TOPLSM 199, **c** the discrete level set method, and **d** the SIMP method

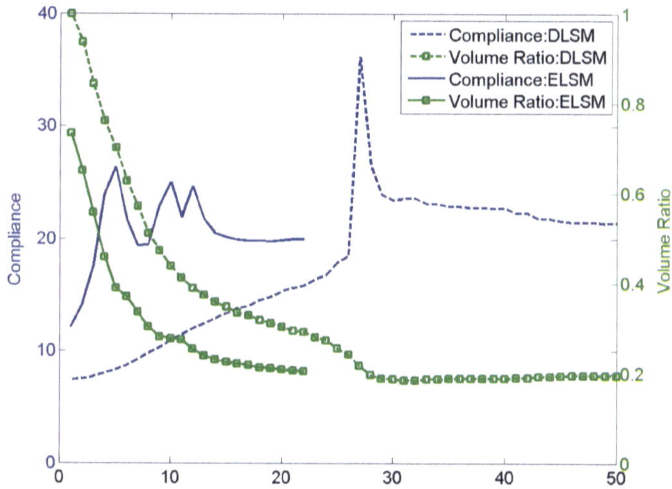

Fig. 5.3 The convergence histories of the structures obtained using the proposed method and the discrete level set method

Table 5.1 Comparison of four approaches to topology optimization in the bridge problem

	J	Vol	t_{FEA}(s)	t(s)	N	T(s)
PM	19.9501	0.2046	2.9551	7.2941	21	153.1761
TOPLSM 199	19.3824	0.2042	2.9560	4.5023	180	810.4140
DLSM	21.3788	1.1956	2.6234	3.1104	50	155.5200
SIMP	22.0884	2.0000	2.9481	3.0313	55	166.7215

Abbreviations: J, objective function (mean compliance); Vol, volume ratio; t_{FEA}, average time in seconds of one FEA; t, average time in seconds of one iteration; N, number of iterations; T, total time in seconds of the optimization process; PM, proposed method; TOPLSM 199, standard level set method; DLSM, discrete level set method; SIMP, solid isotropic microstructure with penalization

In the implementation of the proposed method, although the computation time of each iteration is greater than it is when the SLSM is used (because of the *inner loop*), the overall time can still be decreased considerably. Furthermore, the proposed method with a fine initial guess at the topology requires less computation time than the SIMP method and the discrete level set method.

5.1.3.2 Compliant Mechanism Design Problem

The design domain and the boundary conditions are shown in Fig. 5.4. The ratio of the length and height of design domain is 2 : 1. The design domain is discretized by 80×40 finite elements and therefore, $r_f = 40$ is used. A maximum material usage constraint of 0.3 is considered.

The initial configuration of the displacement inverter is shown in Fig. 5.4. The final configurations of the displacement inverter obtained using the standard level set method and the proposed method are shown in Fig. 5.5. Similar topologies are obtained, which confirms the ability of the proposed method to produce the optimal solution.

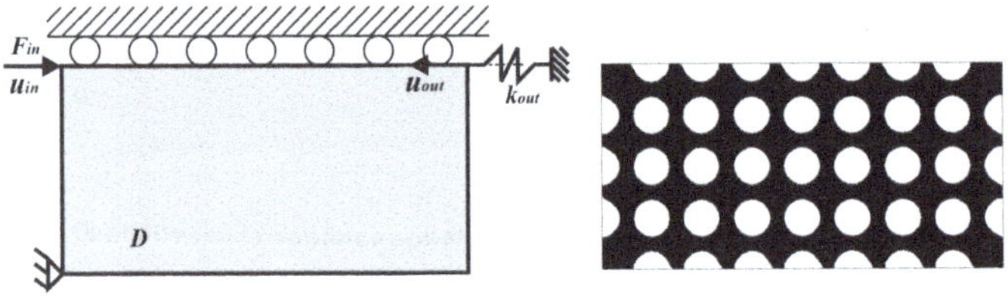

Fig. 5.4 The design domain of the displacement inverter topology optimization problem (left) and the initial configuration (right)

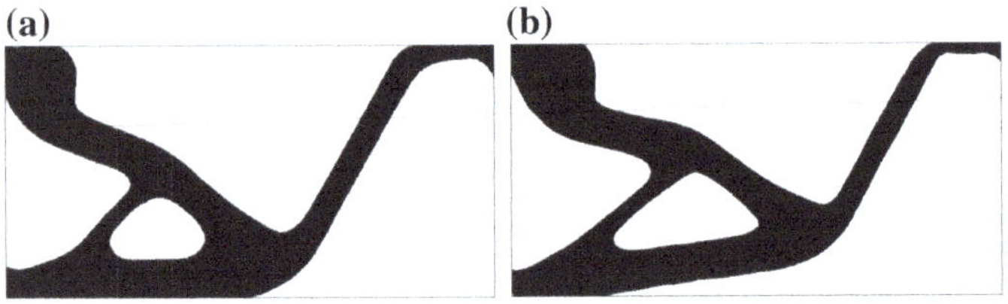

Fig. 5.5 The final designs of a displacement inverter obtained using the conventional level set method (**a**) and the proposed method (**b**)

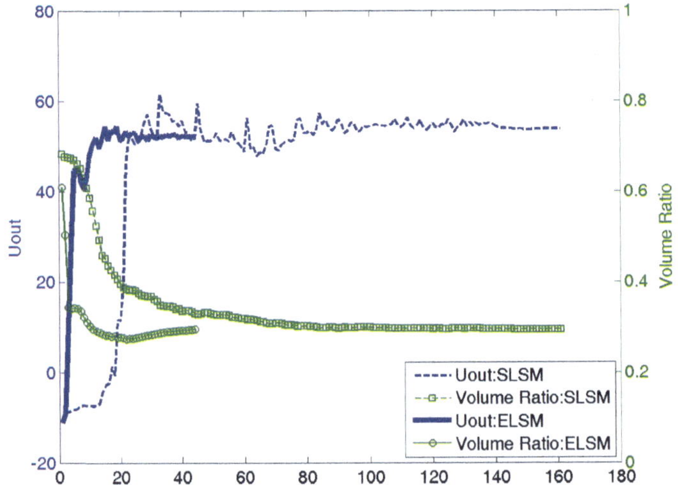

Fig. 5.6 The convergence histories of the displacement inverter problem obtained using the proposed method and the standard level set method

Table 5.2 Comparison of two approaches to the displacement inverter problem

	J	Vol	N	T(s)
SLSM	53.8379	0.2951	161	901.8254
PM	52.0111	0.2958	44	357.3664

Abbreviations: J, objective function (output displacement); Vol, volume ratio; N, total number of iterations; T, total time in seconds of the overall optimization process

The convergence histories are shown in Fig. 5.6. The proposed method requires significantly fewer iterations. Table 5.2 compares the computation time in seconds required to solve the displacement inverter problem using two topology optimization methods. In the implementation of the proposed method, although the computation time required by each iteration is greater than it is when the standard level set method is used, the overall computation time of the optimization process is significantly shorter.

5.2 Topology Optimization of Compliant Mechanisms Considering Geometric Nonlinearity

The topology optimization of compliant mechanisms has been extensively explored using on linear finite element models. However, designing topologies for large-displacement compliant mechanisms using geometrically nonlinear finite element models have not been extensively studied [2, 7, 10, 15, 21]. Here, we present a methodology for the topology optimization of large-displacement compliant mechanisms using the ground structure approach with displacement loading. Frame elements are chosen to represent the design domain. The geometrically nonlinear plane frame structural response is measured using the co-rotational total Lagrangian finite element formulation, and the equilibrium is found using an incremental scheme combined with the Newton-Raphson method.

5.2.1 *Geometrically Nonlinear Finite Element Analysis*

During the nth incremental displacement load, the equilibrium equation for the plane frame element system defined in the total Lagrangian coordinate system (with the superscript n omitted) is

$$\mathbf{R} = \begin{bmatrix} \mathbf{R}^f \\ \mathbf{R}^P \end{bmatrix} = \mathbf{P}_{ext} - \mathbf{F}_{int} = \begin{bmatrix} \mathbf{P}^f \\ \mathbf{P}^P \end{bmatrix} - \begin{bmatrix} \mathbf{F}^f(\mathbf{U}, \mathbf{b}) \\ \mathbf{F}^p(\mathbf{U}, \mathbf{b}) \end{bmatrix} = 0 \tag{5.16}$$

$$\mathbf{U} = \begin{bmatrix} \mathbf{U}^f \\ \mathbf{U}^p \end{bmatrix}, \tag{5.17}$$

where $\mathbf{R}$ is the force residual, $\mathbf{F}_{int}$ are the internal forces, $\mathbf{P}_{ext}$ are the external loads and $\mathbf{U}$ is the displacement response vector. The residuals $\mathbf{R}^f$ and $\mathbf{R}^p$ and the internal forces $\mathbf{F}^f$ and $\mathbf{F}^p$ refer to the points and directions in the structure at which the external loads $\mathbf{P}^f$ and the prescribed displacements $\mathbf{U}^p$ are given. $\mathbf{b}$ denotes the vector of design variables. Rewriting Eq. (5.16) using a first-order Taylor series expansion yields

$$\mathbf{R} + \frac{\partial \mathbf{R}}{\partial \mathbf{U}} d\mathbf{U} = \mathbf{0}. \tag{5.18}$$

From Eq. (5.18), the incremental displacements $\Delta\mathbf{U}$ are obtained by solving the following equation:

$$\mathbf{K}_T \Delta\mathbf{U} = \mathbf{R} = \begin{bmatrix} \mathbf{K}^{ff} & \mathbf{K}^{fp} \\ \mathbf{K}^{pf} & \mathbf{K}^{PP} \end{bmatrix} \begin{bmatrix} \Delta\mathbf{U}^f \\ \Delta\mathbf{U}^p \end{bmatrix} = \begin{bmatrix} \mathbf{R}^f \\ \mathbf{R}^P \end{bmatrix}, \tag{5.19}$$

where $\mathbf{K}_T$, the consistent tangent stiffness, is expressed by

$$\mathbf{K}_T = -\frac{\partial \mathbf{R}}{\partial \mathbf{U}}. \tag{5.20}$$

When only the input displacement $\mathbf{U}^p$ is applied, the structural displacement $\mathbf{U}^f$ is determined using

$$\mathbf{K}^{ff}\left(\mathbf{U}_j^p, \mathbf{U}^p\right) \Delta\mathbf{U}^f = \mathbf{R}^f\left(\mathbf{U}_j^p, \mathbf{U}^p\right) - \mathbf{K}^{fP}\left(\mathbf{U}_j^f, \mathbf{U}^p\right) \Delta\mathbf{U}^p, \tag{5.21}$$

where the subscript j represents the jth Newton-Raphson iteration for the nth incremental displacement load. The displacements are updated using Eqs. (5.19) and (5.21),

$$\begin{aligned} &\mathbf{U}_{j+1}^f = \mathbf{U}_j^f + \Delta\mathbf{U}^f, \\ &j = 1: \quad \Delta\mathbf{U}^p = \Delta\mathbf{U}_n^p, \\ &j \geq 2: \quad \Delta\mathbf{U}^p = 0, \end{aligned} \tag{5.22}$$

where $\Delta\mathbf{U}_n^p$ is the magnitude of the nth incremental displacement load.

5.2.2 Optimization Algorithm

The goal of designing a compliant mechanism is to build a kinematically feasible configuration with flexibility sufficient to obtain the desired output motion and stiffness sufficient to withstand external loads [13, 14]. Therefore, for the given input and output design specifications and the boundary conditions, a multi-objective function that captures the need for both flexibility and stiffness is formulated.

The system flexibility requirement is satisfied by maximizing the geometric advantage to generate the desired kinematic motion in response to the applied input. The geometric advantage is defined by maximizing the ratio of the output displacement U_{out} and the input displacement U_{in} [13]. Since the displacement load is prescribed, this is equivalent to maximizing the output displacement U_{out}. Maximizing the flexibility is formulated as follows:

$$\max \quad GA = \frac{U_{out}}{U_{in}} \quad \Leftrightarrow \quad \max \quad U_{out}. \tag{5.23}$$

The need for the compliant mechanism to be stiff enough to withstand the external load is captured by the stiffness requirement [14]. The rigidity is maximized by minimizing the strain energy. Minimization of the strain energy is formulated as follows:

$$\min \quad SE = \sum_{i=1}^{n} \frac{1}{2} (\Delta \mathbf{U}^i)^T \mathbf{K}_T^i \Delta \mathbf{U}^i, \tag{5.24}$$

where n is the number of the incremental displacement load, $\Delta \mathbf{U}^i$ is the incremental displacement in response to the ith incremental displacement load, $\mathbf{K}_T^i$ is the tangent stiffness when the Newton-Raphson iteration is terminated in response to the ith incremental displacement load.

Therefore, the multi-objective function is formulated as follows:

$$\begin{aligned} \min: & \quad -\frac{U_{out}}{SE} \\ \text{s.t.}: & \quad \mathbf{R} = 0 \\ & \quad V \leq V^* \\ & \quad \mathbf{b}_{low} \leq \mathbf{b} \leq \mathbf{b}_{upp} \end{aligned} , \tag{5.25}$$

where V is the total volume, $V^{\max}$ is the allowed upper bound on the material volume, $\mathbf{b}$ is the in-plane width of the frame elements, namely, the vector of design variables, and $\mathbf{b}_{low}$ and $\mathbf{b}_{upp}$ are the lower and upper bounds on the design variable, respectively.

To apply the gradient-based algorithm, the sensitivity of the objective function with respect to each design variable is needed. Rewriting the equilibrium condition, Eq. (5.16), as function of the design variable vector $\mathbf{b}$ yields

$$\mathbf{R}(\mathbf{U}, \mathbf{b}) = \mathbf{P}_{ext} - \mathbf{F}_{int}(\mathbf{U}, \mathbf{b}) = 0. \tag{5.26}$$

Differentiating the function $\mathbf{R}(\mathbf{U}, \mathbf{b})$ yields

$$\frac{\partial \mathbf{R}(\mathbf{U}, \bar{b})}{\partial \mathbf{U}} \frac{d\mathbf{U}}{d\mathbf{b}} + \frac{\partial \mathbf{R}(\bar{U}, \mathbf{b})}{\partial \mathbf{b}} = 0, \tag{5.27}$$

where a subscript $^-$ means that the variable remains constant during the partial differentiation.

From Eq. (5.27), we obtain

$$\mathbf{K}_T \frac{d\mathbf{U}}{d\mathbf{b}} = -\frac{\partial \mathbf{R}}{\partial \mathbf{b}}. \tag{5.28}$$

Using the adjoint method, the sensitivity of the output displacement U_{out} is expressed as

$$\frac{dU_{out}}{d\mathbf{b}} = -\mathbf{L}_{out}^T \mathbf{K}_T^{-1} \frac{\partial \mathbf{R}}{\partial \mathbf{b}}, \tag{5.29}$$

where $\mathbf{L}_{out}^T$ is chosen to satisfy the following equation:

$$\boldsymbol{\lambda}^T = \mathbf{L}_{out}^T \mathbf{K}_T^{-1}. \tag{5.30}$$

Therefore, Eq. (5.29) becomes

$$\frac{dU_{out}}{d\mathbf{b}} = -\boldsymbol{\lambda}^T \frac{\partial \mathbf{R}}{\partial \mathbf{b}}. \tag{5.31}$$

Differentiating the strain energy SE yields

$$\frac{dSE}{d\mathbf{b}} = \sum_{i=1}^{n} \frac{1}{2}\left(\frac{\partial (\Delta \mathbf{U}^i)^T}{\partial \mathbf{b}} \mathbf{K}_T^i \Delta \mathbf{U}^i + (\Delta \mathbf{U}^i)^T \frac{\partial \mathbf{K}_T^i}{\partial \mathbf{b}} \Delta \mathbf{U}^i + (\Delta \mathbf{U}^i)^T \mathbf{K}_T^i \frac{\partial \mathbf{U}^i}{\partial \mathbf{b}}\right), \tag{5.32}$$

which can be simplified, using the direct derivative of Eq. (5.19), to

$$\frac{dSE}{d\mathbf{b}} = \sum_{i=1}^{n} -\frac{1}{2} (\Delta \mathbf{U}^i)^T \frac{\partial \mathbf{K}_T^i}{\partial \mathbf{b}} \Delta \mathbf{U}^i. \tag{5.33}$$

The sensitivity of the strain energy is not continuous. A simple way to improve the accuracy of the current sensitivity by considering the deformation history of each frame element is to average the current sensitivity with that of the previous iteration [15]. Equation (5.33) can be rewritten as

$$\frac{dSE}{d\mathbf{b}} = \sum_{i=1}^{n} -\frac{1}{4}\left((\Delta \mathbf{U}^{i-1})^T \frac{\partial \mathbf{K}_T^{i-1}}{\partial \mathbf{b}} \Delta \mathbf{U}^{i-1} + (\Delta \mathbf{U}^i)^T \frac{\partial \mathbf{K}_T^i}{\partial \mathbf{b}} \Delta \mathbf{U}^i\right). \tag{5.34}$$

The flowchart of the topology optimization procedure for large-displacement compliant mechanisms is shown in Fig. 5.7. The low-density elements can cause the tangent stiffness matrix to become indefinite, and the Newton-Raphson method fails to converge during the optimization of a continuous ground structure with geometric nonlinearities [15]. In the co-rotating plane frame model, the same problem occurred when the in-plane widths of the frame elements approached the lower bound of the design variable. Saxena et al. [16] periodically removed unstable elements whose

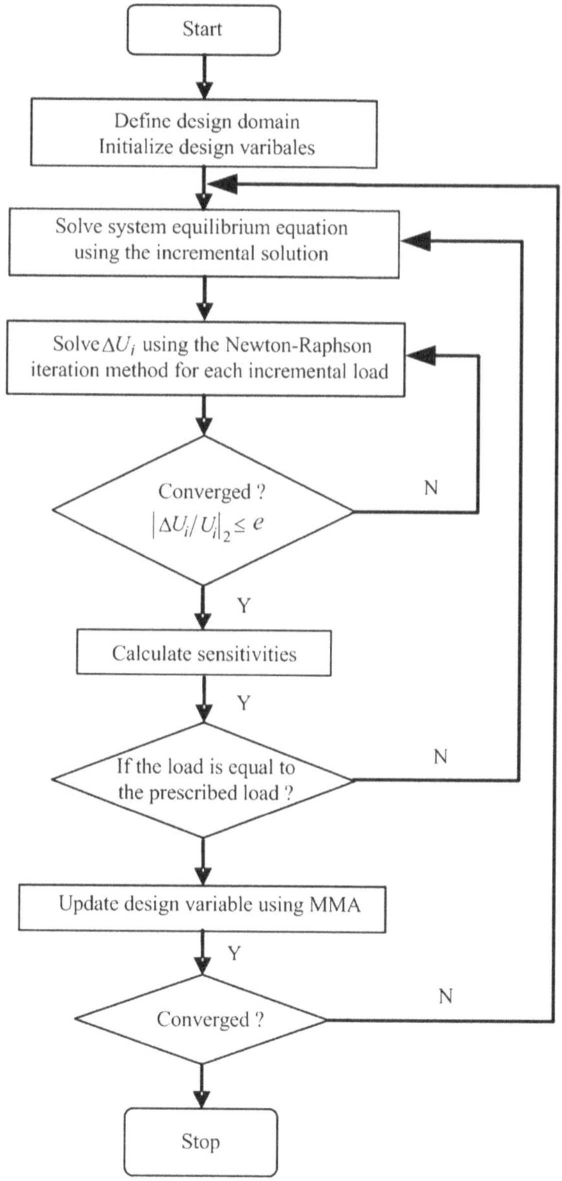

Fig. 5.7 Flowchart of the topology optimization procedure

width approaches the lower bound from the mesh during the optimization process when using the ground structure approach. The method is used here to avoid numerical instability due to unstable elements.

5.2.3 Numerical Examples

Here, we use a classical displacement inverter mechanism to demonstrate the method. Due to the symmetric nature of this problem, only the half of the design domain is used. The size of the design domain is 100 mm by 50 mm. The Young's modulus of the structural material is 169 GPa, and its Poisson ratio is 0.3. An input displacement 0.2 mm is applied and a spring (5 N/mm) is attached to the output port. The lower and upper bounds of the design variable are set to 0.00 mm and 5 mm, respectively. The design domain is divided into 54 frame elements as a reduced ground structure.

The optimal topologies are obtained using geometrically nonlinear theory and linear theory as shown in Fig. 5.8. The optimal topology based on geometrically nonlinear theory is significantly different from the one based on linear theory. For the same input displacement, the output displacement and the strain energy of the optimal mechanism based on geometrically nonlinear theory are 0.196 mm and 280.146 Nmm, respectively. However, the output displacement and the strain energy of the optimal mechanism based on linear theory are 0.158 mm and 312.253 Nmm, respectively. This shows that the optimal mechanism based on geometrically nonlinear theory is stiffer and more flexible than the one based on linear theory.

The resulting data is stored in the IGES format and introduced to ANSYS for the finite element analysis. The output displacement of the optimal mechanism based on geometrically nonlinear theory is 0.192 mm and that of the optimal mechanism based on linear theory is 0.155 mm. The maximum stress of the geometrically nonlinear topology is 129 MPa, and that of the linear topology is 147 MPa. Compared with the latter, the stress concentration is not obvious in the stress distribution of the former. Therefore, the geometrically nonlinear topology distributes stress better than the linear topology. It shows that topological design of compliant mechanisms with geometric nonlinearities is necessary, and the performance of the optimal mechanism based on the geometrically nonlinear formulation is better than that of the optimal mechanism based on the linear formulation.

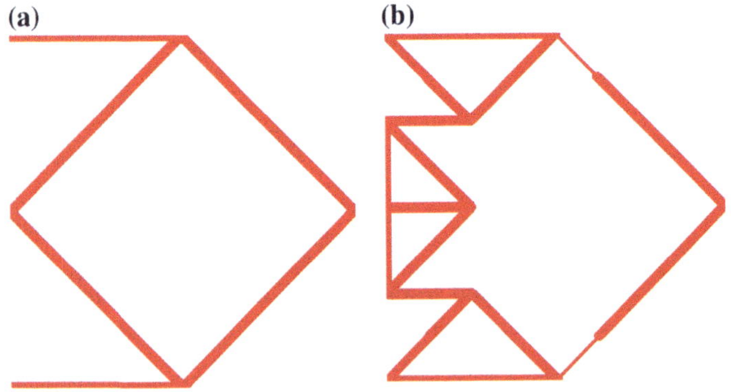

Fig. 5.8 Optimized mechanisms: **a** mechanism based on geometrically nonlinear theory and **b** mechanism based on linear theory

5.3 Reliability-Based Topology Optimization of Compliant Mechanisms

Up-to-date, reliability-based design optimization methods, which allow design for a specific risk and target reliability level, have been frequently applied to the topology optimization of structures [4, 8, 12, 18]. For example, optimizing the topology of reliability-based continuous structures has been considered using first-order reliability methods [8]. A reliability-based design optimization model has been developed for optimizing the topology of structures with material uncertainties [5]. However, only a few approaches combine reliability-based topology optimization methods with the design of compliant mechanisms. Maute and Frangopol [13] proposed a method of designing MEMS using topology optimization that considers stochastic loading and boundary conditions as well as material properties. Allen et al. [1] extended this idea by proposing a method for the reliability-based analysis and design optimization of electrostatically actuated MEMS. Zhang and Ouyang [23] proposed a numerical approach to reliability-based topology optimization for compliant mechanisms with multiple inputs and outputs using the level set method. In that method, a multi-objective model for topology optimization was developed by considering uncertainties in the loads, material properties, and member geometries.

To avoid decreases in the mechanism's performance due to the uncertainty inherent in the production process, a reliability-based topology optimization method for compliant mechanisms is presented. Taking into consideration the randomness of the loads and the geometric description, a mathematical model is for multi-objective optimization of the compliant mechanism. The first-order reliability method (FORM) is adopted to calculate the probabilistic constraint on the reliability index.

5.3.1 *Reliability Analysis*

The inherently probabilistic nature of the design parameters, material properties and loading conditions involved in the structural analysis of a mechanism is an important factor that influences structural safety. Reliability analysis leads to safety measures that a design engineer must take into account due to the aforementioned uncertainty. A time-invariant reliability analysis produces the following relationship:

$$P_f = P[R < S] = \int_{-\infty}^{\infty} f_R(y) f_S(y) dy = 1 = 1 - \int_{-\infty}^{\infty} f_S(y) f_R(y) dy, \tag{5.35}$$

where R denotes the structure's bearing capacity and S represents the external load. The randomness of R and S can be described by known probability density functions $f_R(y)$ and $f_S(y)$, respectively, with $f_R(y) = P[R < y]$ and $f_S(y) = P[S < y]$ being the cumulative probability density functions of R and S, respectively.

Normally, a limit state function is defined as $G(R, S) = S - R$, and the probability of structural failure is given by

$$P_f = P[G(R, S) \geq 0] = \int_{G \geq 0} f_R(R) f_S(S) dR dS. \tag{5.36}$$

A first-order approximation method is generally used to integrate Eq. (5.36). With a first-order approximation of the failure function in the FORM, the failure probability is

$$P_f = \Psi(-\beta), \tag{5.37}$$

where $\Psi(\cdot)$ is the standard normal cumulative distribution function and β is the reliability index. Therefore, by introducing the reliability index β, estimating the failure probability becomes solving for β. In the first-order approximation, the reliability index β is defined as the minimum distance from the origin to most probable point of the limit state surface in the standard normal space S. The most probable point is identified by solving a nonlinear optimization problem with one constraint. For more details, please refer to references [8, 23].

5.3.2 Reliability-Based Topology Optimization

A model for optimizing the topology of a compliant mechanism that considers reliability is as follows [9]:

$$\begin{aligned} \min: &\quad wE_s - (1-w)E_{ms} \\ \text{s.t.}: &\quad \beta(\mathbf{s}) \geq \beta_t \\ &\quad Vol(\mathbf{x}, \mathbf{y}, \mathbf{s}) = Vol^* \end{aligned}, \tag{5.38}$$

where w is the weight, E_s is the strain energy, E_{ms} is the mutual strain energy, Vol is the maximum material usage allowed, which is limited by an upper bound, Vol^*, $\mathbf{x}$ includes the deterministic topology design variables, and $\mathbf{y}$ includes random variables that represent structural uncertainties identified by probabilistic distributions. In this study, $\mathbf{y}$ includes the geometric dimensions, the volume fraction and the applied external load. $\mathbf{s}$ represents normalized variables that also relate their mean values and standard deviations.

The algorithm to solve the optimization problem in the above formulation illustrated in Fig. 5.9. There are two principal iterative loops, reliability analysis and topology optimization. The first loop is part of the FORM. In this loop, the reliability constraint is considered equivalent to a linear function of the random variables. The data from the analysis are processed to obtain random variables that satisfy the reliability constraints. Then, the modified random variables are used for the second loop. The reliability analysis is embedded into the reliability-based topology

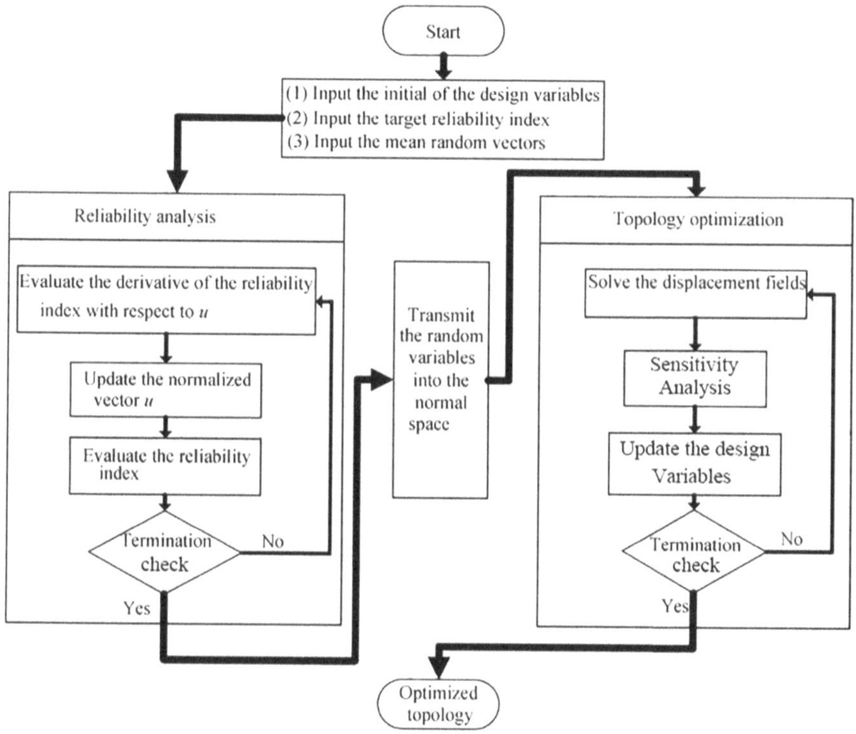

Fig. 5.9 Flow chart for reliability-based topology optimization

optimization framework as an independent loop. This approach avoids performing a reliability analysis during iteration of the topology optimization and decreases the computational time significantly.

5.3.3 Examples

We use a compliant microgripper as an example to show the effect of applying reliability-based topology optimization. As previously mentioned, we consider a square design domain that is discretized into 40×40 bilinear quadrilateral elements. The input load is simply set to 1.

Once the reliability constraints have been satisfied, the values of all the random variables are determined as shown in Table 5.3. We call the topology optimization procedure, and the resulting optimized topology principally depends on the reliability index. Figure 5.10 shows the compliant microgrippers obtained using deterministic and reliability-based topology optimization. The latter method results in a completely different optimal topology than the former method.

Table 5.3 Random variable parameters for $\beta = 5$

Type	Random variable	Initial value	Modified value
Geometry	*nelx*	40	50
	nely	40	30
	Vol	0.4	0.30
Load	*F*	1	1.25

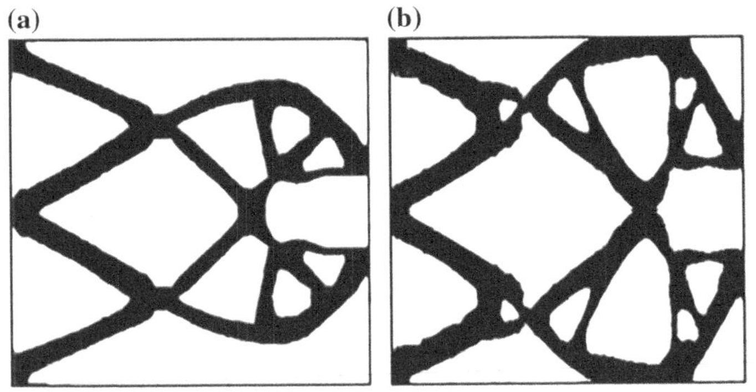

Fig. 5.10 Optimal topology of a compliant microgripper: **a** the optimal deterministic topology and **b** the corresponding reliability-based topology

Another example is presented to illustrate the validity of the design method. The task is to design a micromanipulation stage mechanism with four inputs and four outputs. Figure 5.11 shows the design domain of the compliant mechanism. The input forces are all set to 1, the target reliability index β is 5. The domain is discretized into 40×40 finite elements. The optimized micromanipulation stage mechanisms are illustrated in Fig. 5.12. Figure 5.12a shows the optimal topology when using

Fig. 5.11 Design domain of the micromanipulation stage

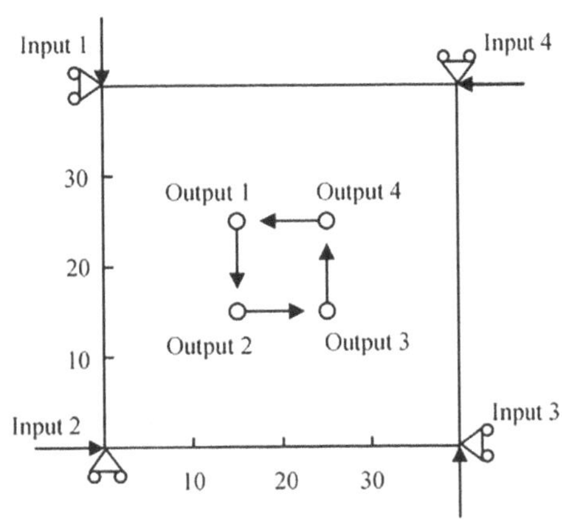

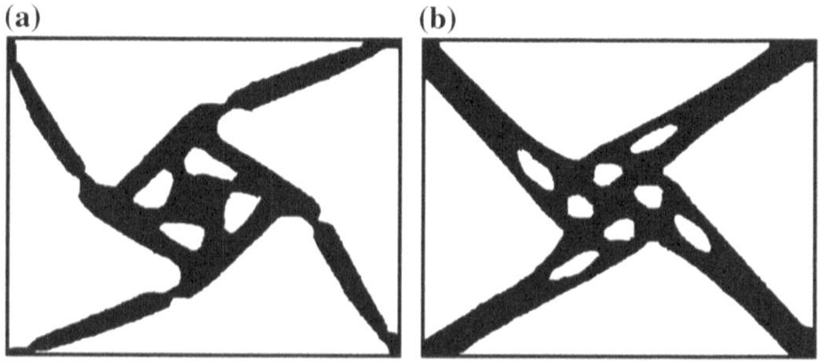

Fig. 5.12 Micromanipulation stage: **a** Deterministic topology and **b** reliability-based topology

the deterministic procedure and Fig. 5.12b presents the topology when reliability is integrated into the optimization procedure.

The above examples show that the strategy of using two iterative loops, reliability analysis and topology optimization, to design the topology of a compliant mechanism is valid and efficient. The randomness of the loads, material properties, and member geometries have a strong effect on the topology optimization results; therefore, it is necessary to consider stochastic nature of the compliant mechanism during the topology optimization process.

5.4 Topology Optimization of Multi-material Compliant Mechanisms

Compared to single-material mechanisms, designing multi-material mechanisms enables designers to fully exploit the advantages of different materials and gives them more design freedom in terms of force, motion, and energy transduction. A new multi-objective topology optimization method for multi-material compliant mechanisms is presented.

5.4.1 Relationship Between Material and Objective

Consider a compliant mechanism with n outputs that is a composite of n materials. Traditionally, the materials and outputs have the relationship shown in Fig. 5.13a, which means that each output behavior is performed by all the materials. There are some difficulties in such multi-material topology optimization problems, e.g., multiple field variables and pointwise constraints on the sum of the volume fraction fields. Here, to release the tight coupling between outputs and materials, we use the mapping relation shown in Fig. 5.13b. In this case, each material corresponds to one specific output. Central to the method is a parallel penalization scheme that works by decomposing the overall multi-material optimization problem into a series of single

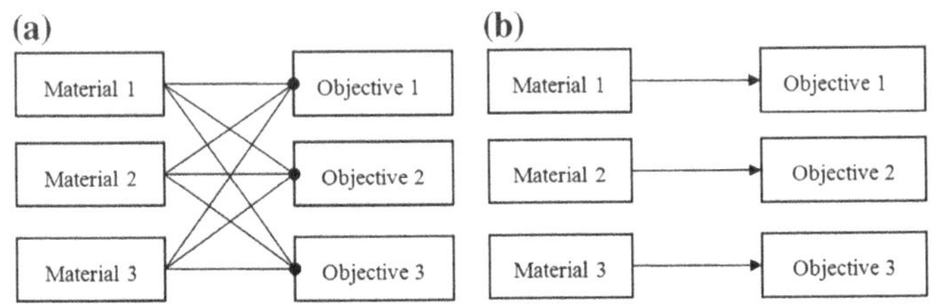

Fig. 5.13 **a** Traditional relationship between materials and outputs and **b** the relationship between materials and outputs used in this chapter

material optimization problems and coordinating the sub-optimizations so that the joint solution is optimal for the overall system. This ensures that each sub-problem contains only one objective [22].

5.4.2 Problem Formulation

In designing compliant mechanisms with multiple materials, the *i*th output displacement is denoted by u_i. Incorporating the SIMP method yields the following optimization model:

$$\begin{aligned} \min_{\mathbf{x}}: \quad & c(\mathbf{x}_i) = -u_i = -\mathbf{U}_{\mathrm{i}}^{\mathrm{T}}\mathbf{K}\mathbf{U}_{\mathrm{in}} \\ \text{s.t.}: \quad & \mathbf{K}\mathbf{U}_{\mathrm{in}} = \mathbf{F}_{\mathrm{in}} \\ & \mathbf{K}\mathbf{U}_{\mathrm{i}} = \mathbf{F}_{\mathrm{i}} \\ & \frac{V(\mathbf{x}_{\mathrm{i}})}{V_0} = f_{\mathrm{i}} \\ & 0 < x_{\mathrm{imin}} \leq \mathbf{x}_{\mathrm{i}} \leq 1 \\ & (\mathrm{i} = 1, 2, \ldots, \mathrm{M}), \end{aligned} \tag{5.39}$$

where $\mathbf{F}_{\mathrm{in}}$ and $\mathbf{U}_{\mathrm{in}}$ are the input load and displacement vectors, respectively, $\mathbf{F}_{\mathrm{i}}$ is a vector with the value 1 at the degree of freedom corresponding to output point *i* and the value zero elsewhere, $\mathbf{U}_{\mathrm{i}}$ is the displacement vector due to $\mathbf{F}_{\mathrm{i}}$, $\mathbf{K}$ is the stiffness matrix, V is the allowable material usage, V_0 is the maximum material volume when the design domain is fully occupied by solid material, f_i is a predefined constant that represents the volume fraction of the *i*th material, and $\mathbf{x}_i$ is the design variable vector corresponding to the *i*th material.

Then, the sensitivity of the objective function is

$$\begin{aligned} \frac{\partial c}{\partial \mathbf{x}_{\mathrm{i}}} &= -\frac{\partial \mathbf{U}_{\mathrm{i}}^{\mathrm{T}}\mathbf{K}\mathbf{U}_{\mathrm{in}}}{\partial \mathbf{x}_{\mathrm{i}}} \\ &= -\left(\frac{\partial \mathbf{U}_{\mathrm{i}}^{\mathrm{T}}}{\partial \mathbf{x}_{\mathrm{i}}}\mathbf{K}\mathbf{U}_{\mathrm{in}} + \mathbf{U}_{\mathrm{i}}^{\mathrm{T}}\frac{\partial \mathbf{K}}{\partial \mathbf{x}_{\mathrm{i}}}\mathbf{U}_{\mathrm{in}} + \mathbf{U}_{\mathrm{i}}^{\mathrm{T}}\mathbf{K}\frac{\partial \mathbf{U}_{\mathrm{in}}}{\partial \mathbf{x}_{\mathrm{i}}}\right). \end{aligned} \tag{5.40}$$

Since $\mathbf{KU}_{\mathrm{in}} = \mathbf{F}_{\mathrm{in}}$ and $\mathbf{KU}_{\mathrm{i}} = \mathbf{F}_{\mathrm{i}}$, we have

$$\mathbf{K}\frac{\partial \mathbf{U}_{\mathrm{in}}}{\partial \mathbf{x}_{\mathrm{i}}} + \frac{\partial \mathbf{K}}{\partial \mathbf{x}_{\mathrm{i}}}\mathbf{U}_{\mathrm{in}} = 0 \tag{5.41}$$

$$\frac{\partial \mathbf{U}_{\mathrm{i}}^{\mathrm{T}}}{\partial \mathbf{x}_{\mathrm{i}}}\mathbf{K} + \mathbf{U}_{\mathrm{i}}^{\mathrm{T}}\frac{\partial \mathbf{K}}{\partial \mathbf{x}_{\mathrm{i}}} = 0. \tag{5.42}$$

Combining all these three equations, we obtain

$$\frac{\partial c}{\partial \mathbf{x}_{\mathrm{i}}} = \mathbf{U}_{\mathrm{i}}^{\mathrm{T}}\frac{\partial \mathbf{K}}{\partial \mathbf{x}_{\mathrm{i}}}\mathbf{U}_{\mathrm{in}} = p(\mathbf{x}_{\mathrm{i}})^{p-1}\mathbf{U}_{\mathrm{i}}^{\mathrm{T}}\mathbf{K}_{\mathrm{i}}\mathbf{U}_{\mathrm{in}}(\mathrm{i} = 1, 2, ..., \mathrm{M}), \tag{5.43}$$

where p is the penalty.

5.4.3 Numerical Implementation

Once the optimization problem for each material has been solved, the design variable $\mathbf{x}_i (i = 1, 2, \ldots, \mathrm{M})$ is obtained. According to the SIMP model, in each element, the density is between 0 and 1. For a multi-material problem, the sum of an element's densities may be greater than 1. This is called an over-saturated element. Obviously, such a unit is contrary to the principle of the model. Therefore, this type of over-saturated element should be treated separately to ensure that the sum of an element's densities is less than or equal to 1. Therefore, the materials are prioritized. The lowest-priority material is removed from an over saturated element first. Suppose that the materials, in order from high to low priority, are $S_1, S_2, \ldots, S_M$ and that the adjusted over-saturated element's density is [22]

$$\mathbf{x}_{S_{(\mathrm{M-k})}}^{e} = \min(0, 1 - \sum_{\mathrm{i=1}}^{\mathrm{M-(k+1)}} \mathbf{x}_{S_{\mathrm{i}}}^{e}) \quad \text{if} \quad \sum_{\mathrm{i=1}}^{\mathrm{M-k}} \mathbf{x}_{S_{\mathrm{i}}}^{e} > 1(\mathrm{k} = 1, 2, ..., \mathrm{M} - 1). \tag{5.44}$$

It is assumed that the properties of the different materials, such as the stiffness tensor, can be superimposed linearly. The stiffness matrix for multiple materials can be calculated using

$$\mathbf{k}(\mathbf{x}_1^e, \mathbf{x}_2^e, \ldots, \mathbf{x}_{\mathrm{M}}^e) = \sum_{\mathrm{i=1}}^{\mathrm{M}} (x_{\mathrm{i}}^e)^p \mathbf{k}_{\mathrm{i}}^e (\mathrm{i} = 1, 2, \ldots, \mathrm{M}, \mathrm{e} = 1, 2, ..., \mathrm{N}), \tag{5.45}$$

where $\mathbf{x}_i^e$ and $\mathbf{k}_i^e$ are the element density and the element stiffness matrix of the ith material, respectively.

5.4.4 Examples

In this section, a multi-objective example based on multi-material topology optimization is presented to illustrate the effectiveness of the previous theoretical model. For the two-material case, the ratio of the elastic modulus of solid material and to that of soft material is assumed to be 10:1 (dark red color: light green color). We use 80×80 bilinear quadrilateral elements for the discretization. The Poisson's ratio ν and the penalty value p are set to 0.3 and 3.5, respectively.

A gripper is considered. Unlike a traditional single output gripper, two lateral outputs are added at the output point. The design domain and the boundary conditions are shown in Fig. 5.14. Because of the symmetry, only half the design domain needs to be considered to obtain a result for the whole mechanism.

In the first case, suppose that the two materials have the same volume fraction, i.e., $V_1 = V_2 = 0.175$. The optimal topology of the gripper is shown in Fig. 5.15a. The optimization process runs for 600 iterations, and the convergence histories of the two output displacements S_{out1} and S_{out2} are shown in Fig. 5.16a.

In the second case, the volume fractions of used two materials are $V_1 = 0.2$ and $V_2 = 0.15$. The optimal topology of the gripper is shown in Fig. 5.15b.

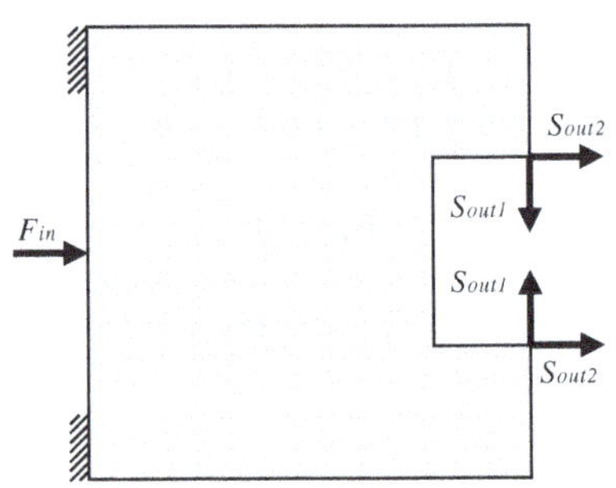

Fig. 5.14 The design domain of a two-material gripper

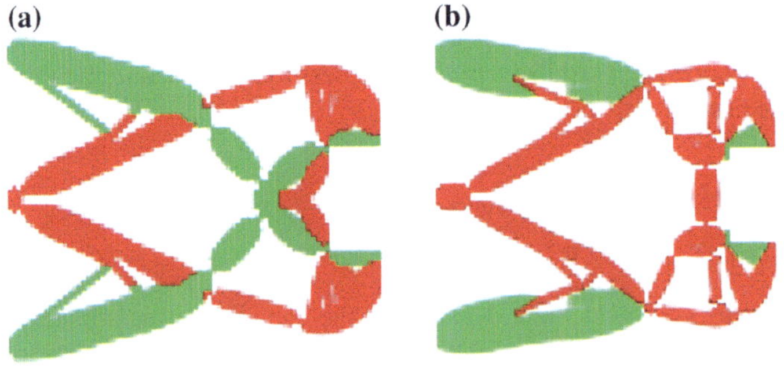

Fig. 5.15 The final design of the gripper: **a** $V_1 = V_2 = 0.175$ and **b** $V_1 = 0.2$; $V_2 = 0.15$. The dark red area corresponds to V_1 and the light green area corresponds to V_2

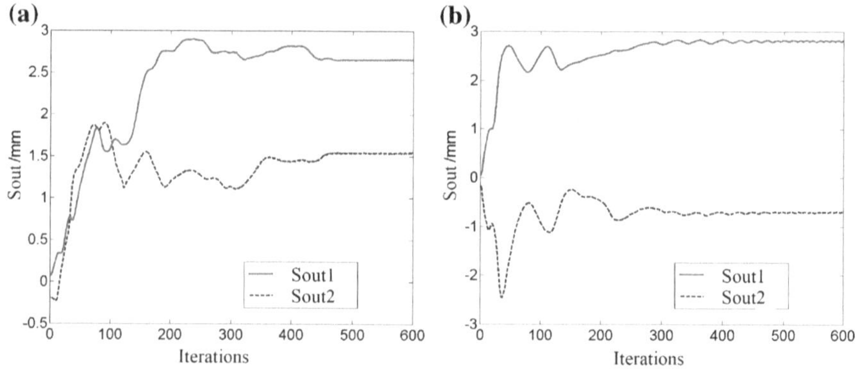

Fig. 5.16 The convergence histories of the two output displacements of the gripper: **a** $V_1 = V_2 = 0.175$ and **b** $V_1 = 0.2; \quad V_2 = 0.15$

The optimization process runs for 600 iterations, and the convergence histories of the two output displacements S_{out1} and S_{out2} are shown in Fig. 5.16b.

The results show that there is a clear boundary between the two materials. From the manufacturing perspective, it can be divided into blocks that are connected by lapping technology. In addition, with the proposed method, one can artificially arrange connections between the material and the design objective.

References

1. Allen, M., Raulli, M., Maute, K., Frangopol, D.M.: Reliability-based analysis and design optimization of electrostatically actuated mems. Comput. Structures **82**(13), 1007–1020 (2004)
2. Bruns, T.E., Tortorelli, D.A.: Topology optimization of non-linear elastic structures and compliant mechanisms. Comput. Methods Appl. Mech. Eng. **190**(26), 3443–3459 (2001)
3. Challis, V.J.: A discrete level-set topology optimization code written in Matlab. Structural Multidisciplinary Optimization **41**(3), 453–464 (2010)
4. Du, J., Sun, C.: Reliability-based vibro-acoustic microstructural topology optimization. Structural Multidisciplinary Optimization **55**(4), 1195–1215 (2017)
5. Jalalpour, M., Tootkaboni, M.: An efficient approach to reliability-based topology optimization for continua under material uncertainty. Structural Multidisciplinary Optimization **53**(4), 759–772 (2016)
6. Jiang, L., Chen, S.: Parametric structural shape & topology optimization with a variational distance-regularized level set method. Comput. Methods Appl. Mech. Eng. **321**, 316–336 (2017)
7. Joo, J., Kota, S.: Topological synthesis of compliant mechanisms using nonlinear beam elements. Mech. Based Des. Structures Mach. **32**(1), 17–38 (2004)
8. Kharmanda, G., Olhoff, N., Mohamed, A., Lemaire, M.: Reliability-based topology optimization. Structural Multidisciplinary Optimization **26**(5), 295–307 (2004)
9. Li, Z., Zhang, X.: Reliability-based topology optimization of compliant micro-gripper with geometrical nonlinearity. J. South China Univ. Technol. (Natural Science Edition) **8**, 023 (2008)
10. Li, Z., Zhang, X.: Topology optimization of multiple inputs and outputs compliant mechanisms with geometrically nonlinearity. Chin. J. Mech. Eng. **45**(1), 180–188 (2009)

11. Luo, J., Luo, Z., Chen, L., Tong, L., Wang, M.Y.: A semi-implicit level set method for structural shape and topology optimization. J. Comput. Phys. **227**(11), 5561–5581 (2008)
12. Luo, Y., Kang, Z., Luo, Z., Li, A.: Continuum topology optimization with non-probabilistic reliability constraints based on multi-ellipsoid convex model. Structural Multidisciplinary Optimization **39**(3), 297–310 (2009)
13. Maute, K., Frangopol, D.M.: Reliability-based design of mems mechanisms by topology optimization. Comput. Struct. **81**(8), 813–824 (2003)
14. Mei, Y., Wang, X.: A level set method for structural topology optimization and its applications. Adv. Eng. Softw. **35**(7), 415–441 (2004)
15. Pedersen, C.B., Buhl, T., Sigmund, O.: Topology synthesis of large-displacement compliant mechanisms. Int. J. Numer. Meth. Eng. **50**(12), 2683–2705 (2001)
16. Saxena, A., Ananthasuresh, G.: Topology synthesis of compliant mechanisms for nonlinear force-deflection and curved path specifications. J. Mech. Des. **123**(1), 33–42 (2001)
17. Sigmund, O.: A 99 line topology optimization code written in Matlab. Structural Multidisciplinary Optimization **21**(2), 120–127 (2001)
18. Silva, M., Tortorelli, D.A., Norato, J.A., Ha, C., Bae, H.R.: Component and system reliability-based topology optimization using a single-loop method. Structural Multidisciplinary Optimization **41**(1), 87–106 (2010)
19. Wang, M.Y., Wang, X., Guo, D.: A level set method for structural topology optimization. Comput. Methods Appl. Mech. Eng. **192**(1), 227–246 (2003)
20. Wang, X., Wang, M., Guo, D.: Structural shape and topology optimization in a level-set-based framework of region representation. Structural Multidisciplinary Optimization **27**(1–2), 1–19 (2004)
21. Zhan, J., Zhang, X.: Topology optimization of compliant mechanisms with geometrical nonlinearities using the ground structure approach. Chin. J. Mech. Eng. **24**(2), 1 (2011)
22. Zhang, X., Hu, K., Wang, N.E.A.: Multi-objective topology optimization of multiple materials compliant mechanisms based on parallel strategy. J. Mech. Eng. (in Chinese) **52**, 1–8 (2016)
23. Zhang, X., Ouyang, G.: A level set method for reliability-based topology optimization of compliant mechanisms. Sci. China Ser. E Technol. Sci. **51**(4), 443–455 (2008)
24. Zhou, M., Wang, M.Y.: A semi-Lagrangian level set method for structural optimization. Structural Multidisciplinary Optimization **46**(4), 487–501 (2012)
25. Zhu, B., Zhang, X.: A new level set method for topology optimization of distributed compliant mechanisms. Int. J. Numer. Meth. Eng. **91**(8), 843–871 (2012)

Erratum to: Topology Optimization of Compliant Mechanisms

Erratum to:
X. Zhang and B. Zhu, *Topology Optimization of Compliant Mechanisms*, https://doi.org/10.1007/978-981-13-0432-3

In the original version of the book, the incorrect first author name "Xianming Zhang" has to be corrected to read as "Xianmin Zhang" in Frontmatter and Cover. The erratum book has been updated with the change.

The updated online version of this book can be found at
https://doi.org/10.1007/978-981-13-0432-3

X. Zhang and B. Zhu, *Topology Optimization of Compliant Mechanisms*,
https://doi.org/10.1007/978-981-13-0432-3_6

Appendix

A.1 Proof in Chap. 5

Proposition A.1 *Assume that V_{n1} and V_{n2} are obtained from a shape sensitivity analysis and that each guarantees a decrease in the Lagrangian L. The velocity field obtained V_{n3} using Eqs.* (5.8) *and* (5.9) *guarantees a decrease in the Lagrangian L.*

Proof (Sketch) In the third step, the level set function is updated based on the velocity field V_{n3}, i.e.,

$$\phi_4 = \phi_3 + V_{n3}\Delta t \tag{A.1}$$

and the Lagrangian's value is indicated by $L(\phi_4)$. Therefore,

$$L(\phi_4) - L(\phi_3) = L(\phi_3 + V_{n3}\Delta t) - L(\phi_3). \tag{A.2}$$

Substituting Eqs. (5.5) and (A.1) into Eq. (A.2) yields

$$\begin{aligned} L(\phi_4) - L(\phi_3) =& L(\phi_2 + (1+\alpha)V_{n2}\Delta t + (1-\alpha)V_{n1}\Delta t) \\ & - L(\phi_2 + V_{n2}\Delta t). \end{aligned} \tag{A.3}$$

Using the Taylor series expansion of Eq. (A.3) yields

$$\begin{aligned} & L(\phi_4) - L(\phi_3) \\ & \doteq L(\phi_2) + (\left.\frac{\partial L(\phi)}{\partial \phi}\right|_{\phi=\phi_2})^T((1+\alpha)V_{n2}\Delta t + (1-\alpha)V_{n1}\Delta t) \\ & \quad - L(\phi_2) - (\left.\frac{\partial L(\phi)}{\partial \phi}\right|_{\phi=\phi_2})^T(V_{n2}\Delta t) \\ & = (\left.\frac{\partial L(\phi)}{\partial \phi}\right|_{\phi=\phi_2})^T(\alpha V_{n2}\Delta t + (1-\alpha)V_{n1}\Delta t). \end{aligned} \tag{A.4}$$

X. Zhang and B. Zhu, *Topology Optimization of Compliant Mechanisms*,
https://doi.org/10.1007/978-981-13-0432-3

Note that since Δt is small, all terms of Eq. (A.4) that contain $(\Delta t)^m (m \geq 2)$ are neglected. Substituting Eq. (5.7) into Eq. (A.4) yields

$$\begin{aligned} L(\phi_4) - L(\phi_3) &= -V_{n2}^T(\alpha V_{n2} + (1-\alpha)V_{n1})\Delta t \\ &= (-\alpha|V_{n2}|^2 - (1-\alpha)V_{n2}^T V_{n1})\Delta t. \end{aligned} \tag{A.5}$$

Since both V_{n2} and V_{n1} guarantee decreases in the Lagrangian L, we have $V_{n2}^T V_{n1} > 0$. Substituting Eq. (5.9) into Eq. (A.5) yields

$$\begin{aligned} L(\phi_4) - L(\phi_3) &= (-\frac{{V_{n2}}^T V_{n1}}{1 + {V_{n2}}^T V_{n1}}|V_{n2}|^2 - \frac{(V_{n2})^T V_{n1}}{1 + {V_{n2}}^T V_{n1}})\Delta t \\ &= -\frac{{V_{n2}}^T V_{n1}(|V_{n2}|^2 + 1)}{1 + {V_{n2}}^T V_{n1}}\Delta t < 0. \end{aligned} \tag{A.6}$$

Proposition A.2 *Even though* $V_{n3} \neq -\left.\frac{\partial L(\phi)}{\partial \phi}\right|_{\phi=\phi_3}$, *the choice of* α *according to Eq.* (5.9) *guarantees a decrease in the Lagrangian* $L(\phi_5)$ *compared to* $L(\phi_4)$, *where*

$$\phi_5 = \phi_4 + V_{n4}\Delta t \tag{A.7}$$

$$V_{n4} = \alpha V_{n3} + (1-\alpha)V_{n2} \tag{A.8}$$

$$\alpha = \frac{V_{n3}^T V_{n2}}{1 + V_{n3}^T V_{n2}}. \tag{A.9}$$

Proof (Sketch)

$$L(\phi_5) - L(\phi_4) = L(\phi_4 + V_{n4}\Delta t) - L(\phi_4) \tag{A.10}$$

Substituting Eqs. (5.5), (A.1) and (A.8) into Eq. (A.10) yields

$$\begin{aligned} L(\phi_5) - L(\phi_4) =& L(\phi_2 + (1+\alpha)V_{n3}\Delta t + (2-\alpha)V_{n2}\Delta t) \\ &- L(\phi_2 + V_{n2}\Delta t + V_{n3}\Delta t). \end{aligned} \tag{A.11}$$

Using the Taylor series expansion of Eq. (A.11) yields

$$\begin{aligned} &L(\phi_5) - L(\phi_4) \\ &\doteq L(\phi_2) + (\left.\frac{\partial L(\phi)}{\partial \phi}\right|_{\phi=\phi_2})^T((1+\alpha)V_{n3}\Delta t + (2-\alpha)V_{n2}\Delta t) \\ &\quad - L(\phi_2) - (\left.\frac{\partial L(\phi)}{\partial \phi}\right|_{\phi=\phi_2})^T(V_{n2}\Delta t + V_{n3}\Delta t) \\ &= (\left.\frac{\partial L(\phi)}{\partial \phi}\right|_{\phi=\phi_2})^T(\alpha V_{n3}\Delta t + (1-\alpha)V_{n2}\Delta t) \\ &= -V_{n2}^T(\alpha V_{n3} + (1-\alpha)V_{n2})\Delta t. \end{aligned} \tag{A.12}$$

Note that since Δt is small, all terms of Eq. (A.12) that contain $(\Delta t)^m (m \geq 2)$ are neglected. Since both V_{n3} and V_{n2} guarantee decreases in the Lagrangian L, we have $V_{n3}^T V_{n2} > 0$. Substituting Eq. (A.9) into Eq. (A.12) yields

$$\begin{aligned} L(\phi_5) - L(\phi_4) &= -\alpha V_{n3}^T V_{n2} \Delta t - (1-\alpha)|V_{n2}|^2 \Delta t \\ &= -\frac{|V_{n3}^T V_{n2}|^2}{1 + V_{n3}^T V_{n2}} \Delta t - \frac{|V_{n2}|^2}{1 + V_{n3}^T V_{n2}} \Delta t < 0. \end{aligned} \tag{A.13}$$

A.2 A 169 Line Matlab Code

This appendix contains the 169-line MATLAB program for optimizing the topologies of hinge-free compliant mechanisms based on weighted-sum method with self-adjusting weights discussed in Chap. 3. This program is partially inspired by the code shared by Prof. Yu. M. Wang at http://www2.mae.cuhk.edu.hk/~cmdl/down load.htm. The 169 lines comprise 35 lines for the main program, 19 lines for the initial configuration, 36 lines for constructing the velocity, 12 lines for updating the level set function, 12 lines for re-initializing the level set function and 55 lines for the finite element analysis.

A.2.1 Matlab Implementation

The main program is called from the MATLAB prompt using

```
Hf_CM(width, height, nelx, nely, maxloop, l_vol)
```

where `width` and `height` are the width and height of the design domain, respectively, `nelx` and `nely` are the numbers of elements in the horizontal and vertical directions, respectively, `maxloop` is the maximum number of iterations allowed and `l_vol` is the Lagrange multiplier for the volume constraint. Other variables and the boundary conditions for the compliant mechanism are defined in the MATLAB code itself. One can edit the relevant lines if needed. In each iteration, MATLAB graphs the current topology. Figure A.1 shows the topology of a displacement inverter obtained by the code when called with the input line

```
Hf_CM(2, 1, 80, 40, 300, 0.45)
```

Re-initialization of the level set function to a signed distance function is important during the optimization process because it ensures stability. However, there is no need to re-initialize the level set function frequently. In the 169-line MATLAB program, the frequency of re-initialization is controlled on line 26. Re-initialization is performed every five steps. The user can specify the number of steps if necessary.

The function `level_set_equation` solves the level set equation so that a new structural topology can be obtained. The time-step is chosen according to the CFL condition and is normally very small. Therefore, one may update the level set

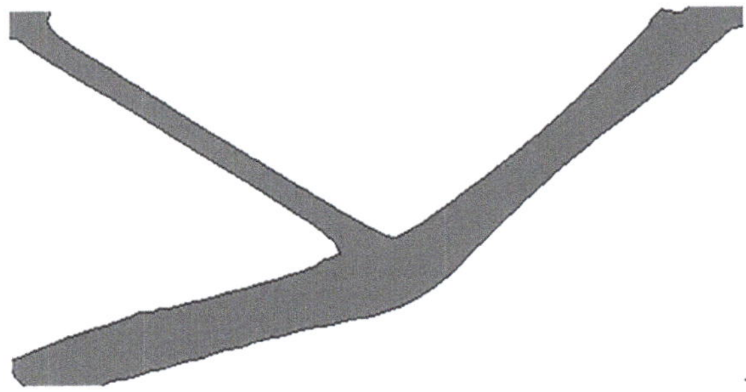

Fig. A.1 The hinge-free compliant displacement inverter obtained using the 199-line MATLAB program

function several times using the same velocity field, as indicated in lines 96-102. The number of time steps is controlled by specifying the value of `it`.

A.2.2 Matlab Code

The following is a transcript of the Matlab code:

```
1  % A 169 LINE MATLAB CODE FOR TOPOLOGY OPTIMIZATION
2  % OF COMPLIANT MECHANISMS USING A WEIGHTTING SUM
3  % METHOD WITH SELF-ADJUSTED WEIGHTING FACTORS
4  function Hf_CM(width, hight, nelx, nely,maxloop,l_vol)
5  %% PARAMETERS INITIALIZATION
6  E1=1; E0 = 1e-3; nu = 0.3;
7  ew = width/nelx; eh = hight/nely;
8  [X,Y] = meshgrid(ew*[-0.5:nelx+0.5],eh*[-0.5:nely+0.5]);
9  LSgrid_x = X(:); LSgrid_y = Y(:);
10 [M,N] = meshgrid(ew*[0:nelx],eh*[0:nely]);
11 FENgrid_x = M(:); FENgrid_y = N(:);
12 %% INITIAL CONFIGUATION
13 phi = init_config(width, hight, nely, nelx, ew, eh, ...
       LSgrid_x, LSgrid_y);
14 phi_fea = griddata(LSgrid_x, LSgrid_y, phi(:), FENgrid_x, ...
       ENgrid_y, 'cubic');
15 loop = 1;
16 alp = 0; bta = 0;
17 %% MAIN OPTIMIZATION LOOP
18 while loop ≤ maxloop
19     %% finite element analysis
20     [U,V,C1,C2,Uout] = FEA(nely, nelx, E1, E0, ew, eh, ...
           nu, reshape(phi_fea,nely+1,nelx+1));
21     %% construct the velocity
22     vel = velocity(E1, E0, U, ew, eh, nely, nelx, l_vol, ...
           alp, bta, phi, nu);
23     %% update the level set equation
24     phi = level_set_equation(phi, ew, eh, vel);
25     %% re-initialize the level set function to signed ...
           distance function
26     if mod(loop,5)==0; phi = reinit_SDF(phi, ew, eh); end;
27     %% calculate the weighting factors of Cin and Cout
```

```
    alp = abs(Uout/C1); bta = abs(Uout/C2);
    %% print results
    disp(['It.:' sprintf('%4i',loop) 'Uout:' ...
        sprintf('%10.2f',full(Uout)) 'V_Ratio:' ...
        sprintf('%10.2f',V/(width*hight)) ])
    %% plot topologies
    phi_fea = griddata(LSgrid_x,LSgrid_y,phi(:), ...
        FENgrid_x,FENgrid_y,'cubic');
    colormap(gray); ...
        contourf(reshape(phi_fea,nely+1,nelx+1),[0 0]); ...
        axis equal; axis tight; axis off; pause(1e-6);
    loop = loop+1;
end
%% SUBFUNCTIONS
function [phi] = init_config(width, hight, nely, nelx, ...
    ew, eh, LSgrid_x, LSgrid_y)
% Preset an initial configuration with several holes
cx = width*[  2/16 4/16 6/16 8/16 10/16 12/16 14/16    ...
            0 2/16 4/16 6/16 8/16 10/16 12/16 14/16 1 ...
            0 2/16 4/16 6/16 8/16 10/16 12/16 14/16 1 ...
            0 2/16 4/16 6/16 8/16 10/16 12/16 14/16 1 ...
              2/16 4/16 6/16 8/16 10/16 12/16 14/16];
cy = hight*[      0    0    0    0    0    0    0      ...
            1/4 1/4 1/4 1/4 1/4 1/4 1/4 1/4 1/4 ...
            2/4 2/4 2/4 2/4 2/4 2/4 2/4 2/4 2/4 ...
            3/4 3/4 3/4 3/4 3/4 3/4 3/4 3/4 3/4 ...
                1    1    1    1    1    1    1];
for i = 1:length(cx)
    tmpPhi(:,i) = -sqrt((LSgrid_x-cx(i)).^2 + ...
        (LSgrid_y-cy(i)).^2) + hight/11;
end
TmpPhi = -(max(tmpPhi.')).';
phi = reshape(TmpPhi,nely+2,nelx+2);
phi = reinit_SDF(phi,ew,eh);
function [velocity] = velocity(E1, E0, U, Ew, Eh, nely, ...
    nelx, L_Vol, c1, c2, phi, nu)
% Construct the velocity field for updating the
% level set function phi
phi = phi(2:nely+1,2:nelx+1);
Dsub = (1-nu^2)*[1,    nu,      0
                 nu,   1,       0
                 0,    0,   (1-nu)/2];
B = 1/2*[-1/Ew   0    1/Ew   0    1/Ew   0   -1/Ew   0
            0   -1/Eh   0   -1/Eh   0   1/Eh   0   1/Eh
         -1/Eh -1/Ew -1/Eh  1/Ew 1/Eh 1/Ew 1/Eh -1/Ew];
for ely = 1:nely
    for elx = 1:nelx
        n1 = (nely+1)*(elx-1)+ely;
        n2 = (nely+1)*elx    +ely;
        Ue1 = U([2*n1-1; 2*n1; 2*n2-1; 2*n2; 2*n2+1; ...
            2*n2+2; 2*n1+1; 2*n1+2],1);
        Ue2 = U([2*n1-1; 2*n1; 2*n2-1; 2*n2; 2*n2+1; ...
            2*n2+2; 2*n1+1; 2*n1+2],2);
        Stain1 = B*Ue1;
        Stain2 = B*Ue2;
        if phi(ely,elx) > 0.75*Ew
            ERatio = E1;
        elseif phi(ely,elx) < -0.75*Ew
            ERatio = E0;
```

```
        else
            Value = phi(ely,elx)/(0.75*Ew);
            ERatio = E1*0.75*(1 - 1e-3)*(Value - ...
                Value^3/3) + 0.5*(1+1e-3);
        end
        D = ERatio*Dsub;
        Beta_u1_out(ely,elx) = Stain2'*D*Stain1;
        Beta_u1_in(ely,elx)= Stain1'*D*Stain1;
        Beta_u2_out(ely,elx)= Stain2'*D*Stain2;
    end
end
Beta = L_Vol - c1 * Beta_u1_in - c2 * Beta_u2_out + ...
    Beta_u1_out;
velocity(1:nely+2,1:nelx+2) = 0;
velocity(2:nely+1,2:nelx+1) = Beta/max(max(abs(Beta)));
velocity = velocity .* exp(1-abs(velocity));
function [phi]=level_set_equation(phi, dx, dy, Vn)
% Solve the level set equation using the ENO2 scheme
Vn_ext = init_normal_ENO2(Vn);
it = 0;
t = 0;
while (it < 5)
    [delt_normal,H1_abs,H2_abs] = evolve_normal_ENO2(phi, ...
        dx, dy, Vn_ext);
    dt = get_dt_normal(0.5,dx,dy,H1_abs,H2_abs);
    phi = phi-dt*delt_normal;
    it = it+1;
    t = t+dt;
end
function [phi] = reinit_SDF(phi, dx, dy)
% Re-initialize the level set function
% to be a signed distance function
s_phi_0 = phi./sqrt(phi.^2+dx.^2);
Vn_ext = init_normal_ENO2(s_phi_0);
it=0;
while (it < 30)
    [delt_normal,H1_abs,H2_abs] = evolve_normal_ENO2(phi, ...
        dx, dy, Vn_ext);
    dt = get_dt_normal(0.5,dx,dy,H1_abs,H2_abs);
    phi = phi+dt*(s_phi_0-delt_normal);
    it = it+1;
end
function [U,V,u1_in,u2_out,u1_out] = FEA(nely, nelx, E1, ...
    E0, ew, eh, nu, phi)
% Finite element analysis
K = sparse(2*(nely+1)*(nelx+1), 2*(nely+1)*(nelx+1));
F = sparse(2*(nely+1)*(nelx+1), 2);
U = sparse(2*(nely+1)*(nelx+1), 2);
AreaFactor_V(1:nely,1:nelx) = 0;
for ely = 1:nely
    for elx = 1:nelx
        [KE,AreaFactor] = Real_ES(E1, E0, ew, eh, nu, ...
            phi(ely:ely+1,elx:elx+1));
        AreaFactor_V(ely,elx) = AreaFactor;
        n1 = (nely+1)*(elx-1)+ely;
        n2 = (nely+1)*elx     +ely;
        edof = [2*n1-1; 2*n1; 2*n2-1; 2*n2; 2*n2+1; ...
            2*n2+2; 2*n1+1; 2*n1+2];
        K(edof,edof) = K(edof,edof) + KE;
```

```
        end
    end
    din = 2*(nely+1)-1;
    dout = 2*(nely+1)*(nelx+1)-1;
    F(din,1) = 0.1;
    F(dout,2) = -0.1;
    fixeddofs = ...
        union([2*(nely+1):2*(nely+1):2*(nely+1)*(nelx+1)],[1:1:4]);
    alldofs = [1:2*(nely+1)*(nelx+1)];
    freedofs = setdiff(alldofs, fixeddofs);
    U(freedofs,:) = K(freedofs, freedofs) \ F(freedofs,:);
    U(fixeddofs,:)= 0;
    u1_in = U(din, 1);
    u1_out = U(dout, 1);
    u2_out = U(dout, 2);
    V = sum(sum(AreaFactor_V))*ew*eh;
    function [RealKE,AreaFactor]=Real_ES(E1, E0, Ew, Eh, nu, phi)
    % Return the real element stiffness matrix
    if min(min(phi)) > 0
        E = E1; AreaFactor = 1;
    elseif max(max(phi)) < 0
        E = E0; AreaFactor = 0;
    else
        [s,t] = meshgrid([-1:0.1:1],[-1:0.1:1]);
        tmpphi = (1-s(:)).*(1-t(:))/4*phi(1,1) + ...
            (1+s(:)).*(1-t(:))/4*phi(1,2) + ...
            (1+s(:)).*(1+t(:))/4*phi(2,1) + ...
            (1-s(:)).*(1+t(:))/4*phi(2,2);
        E = length(find(tmpphi≥0))/length(s(:))*E1;
        AreaFactor = length(find(tmpphi≥0))/length(s(:));
    end
    RealKE = Basic_ES(E,nu,Ew,Eh);
    function [KE] = Basic_ES(E, nu, a, b)
    % Basic element stiffness matrix
    k = [   -1/6/a/b*(nu*a^2-2*b^2-a^2),     1/8*nu+1/8, ...
        -1/12/a/b*(nu*a^2+4*b^2-a^2),       3/8*nu-1/8, ...
        1/12/a/b*(nu*a^2-2*b^2-a^2),       -1/8*nu-1/8,       ...
            1/6/a/b*(nu*a^2+b^2-a^2),        -3/8*nu+1/8];
    KE = E/(1-nu^2)*...
                        [ k(1) k(2) k(3) k(4) k(5) k(6) k(7) k(8)
                          k(2) k(1) k(8) k(7) k(6) k(5) k(4) k(3)
                          k(3) k(8) k(1) k(6) k(7) k(4) k(5) k(2)
                          k(4) k(7) k(6) k(1) k(8) k(3) k(2) k(5)
                          k(5) k(6) k(7) k(8) k(1) k(2) k(3) k(4)
                          k(6) k(5) k(4) k(3) k(2) k(1) k(8) k(7)
                          k(7) k(4) k(5) k(2) k(3) k(8) k(1) k(6)
                          k(8) k(3) k(2) k(5) k(4) k(7) k(6) k(1)];
```

It should be pointed out that several subfunctions, i.e., `select_der_normal`, `init_normal_ENO2`, `evolve_normal_ENO2`, `get_dt_normal`, `der_ENO2_minus` and `der_ENO2_plus`, can be freely downloaded directly from the following website: http://vision.ece.ucsb.edu/~sumengen/.

Although ENO2 is used to calculate the derivative of the level set function ϕ, other schemes with different accuracies can be used in a straightforward way. One can use ENO1, ENO3 or WENO, which correspond to 1st-, 3rd- and 5th-order schemes, to calculate the derivative of ϕ by simply replacing the corresponding subfunctions. For example, one can use `init_normal_ENO3`, `evolve_normal_ENO3`, `der_ENO3_minus` or `der_ENO3_plus` instead of `init_normal_ENO2`, `evolve_normal_ENO2`, `der_ENO2_minus` and `der_ENO2_plus` for 3rd-order accuracy.

The manufacturer's authorised representative in the EU is Springer Nature Customer Service Centre GmbH, Europaplatz 3, 69115 Heidelberg, Germany. If you have any concerns regarding our products, please contact ProductSafety@springernature.com

Printed and bound by CPI Group (UK) Ltd, Croydon, CR0 4YY
25/09/2026
02222891-0001